Perspectives in Alcohol and Drug Abuse

Similarities and Differences

Edited by

Joel Solomon
Kim A. Keeley

John Wright • PSG Inc
Boston Bristol London
1982

Library of Congress Cataloging in Publication Data
Main entry under title:

Perspectives in alcohol and drug abuse.

Bibliography: p.
Includes index.
Contents: A historical review of drug and alcohol use / Benjamin Kissin -- Sociocultural aspects of alcohol and drug use and abuse / Joseph Westermeyer -- The pharmacology of addictive drugs / Kenneth Blum, Arthur H. Briggs, Karl Verebey -- [etc.]
1. Alcoholism--Addresses, essays, lectures.
2. Drug abuse--Addresses, essays, lectures.
I. Solomon, Joel. II. Keeley, Kim A. [DNLM:

1. Substance dependence. WM 270 S689p]
HV5035.P44 362.2'9 81-15918
ISBN 0-88416-306-7 AACR2

Published by:
John Wright • PSG Inc, 545 Great Road, Littleton, Massachusetts 01460, U.S.A.
John Wright & Sons Ltd, 42–44 Triangle West, Bristol BS8 1EX, England

Printed in Great Britain by
John Wright & Sons (Printing) Ltd. at The Stonebridge Press, Bristol.

International Standard Book Number: 088416-306-7

Library of Congress Catalog Card Number: 81-15918

CONTRIBUTORS

Kenneth Blum, PhD
Chief, Division of Substance and Alcohol Misuse
Health Science Center at San Antonio
University of Texas
San Antonio, Texas

Arthur H. Briggs, MD
Division of Substance and Alcohol Misuse
Health Science Center at San Antonio
University of Texas
San Antonio, Texas

Jerome F.X. Carroll, PhD
Director, Psychological Services
Eagleville Hospital and Rehabilitation Center
Eagleville, Pennsylvania

Sidney Cohen, MD
Clinical Professor of Psychiatry
Neuropsychiatric Institute
University of California at Los Angeles
Los Angeles, California

Douglas A. Eldridge, Esq
Counsel
New York State Division of Substance Abuse Services
New York, New York

Bradley D. Evans, MD
Assistant Professor of Psychiatry and Human Behavior
Thomas Jefferson University School of Medicine
Philadelphia, Pennsylvania

Edward Gottheil, MD, PhD
Professor of Psychiatry and Human Behavior
Thomas Jefferson University School of Medicine
Philadelphia, Pennsylvania

Kim A. Keeley, MD, MSH
Clinical Associate Professor
Department of Psychiatry
State University of New York
Downstate Medical Center
Brooklyn, New York

Benjamin Kissin, MD
Director, Division of Alcoholism and Drug Dependence
State University of New York
Downstate Medical Center
Brooklyn, New York

Donald J. Ottenberg, MD
Director
Eagleville Hospital and Rehabilitation Center
Eagleville, Pennsylvania

Joel Solomon, MD
Clinical Associate Professor
Department of Psychiatry
State University of New York
Downstate Medical Center
Brooklyn, New York

John D. Swisher, PhD
Professor of Education
Director, Center for Research on Human Resources
The Institute for Policy Research and Evaluation
The Pennsylvania State University
State College, Pennsylvania

Karl Verebey, MD
Chief, Clinical Pharmacology
Bureau of Laboratory and Testing
New York State Division
of Substance Abuse Services
Brooklyn, New York

Judith R. Vicary, PhD
Assistant Professor of
Administration of Justice
The Pennsylvania State University
State College, Pennsylvania

Joseph Westermeyer, MD, MPH, PhD
Professor
Department of Psychiatry
University of Minnesota
Medical School
Minneapolis, Minnesota

CONTENTS

INTRODUCTION

The use, misuse, and abuse of psychoactive chemicals in one form or another has been present in almost all societies since the beginning of recorded civilization. These chemicals, called (among other things) spirits, tonics, medication, drugs, or substances of abuse, have been widely used in a variety of religious, experimental, medical, and social contexts. By and large, they have been used effectively and accomplished the purpose for which they were intended, for example, the relief of pain or anxiety; the induction of relaxation and sleep; an intoxication euphoria; or the development of an altered state of consciousness for religious or experimental purposes.

Since the turn of the century, and especially within the past 25 years, there has been a rapid proliferation of synthetic and semisynthetic psychoactive substances used, the numbers of people using them, and the variety of usage patterns, particularly multiple or polydrug use and abuse. All of this has led to a phenomenology of chemical use that is particularly complex. Attempts to simplify it through unifying concepts or sweeping generalizations tend to increase the misconceptions and confusion.

Inconsistency and lack of clarity also exist in the use of terminology. For example, drug abuse for some people means only opiate addiction, whereas for others it is opiate addiction and the use of so-called "soft" drugs. For still others, it encompasses any substance that has abuse potential, including alcohol. For many who would like to see a unification of programs and services, the generic concept of chemical dependence or substance abuse has replaced the drug-specific concept of alcoholism or opiate addiction. For others, the one-drug-one-problem concept has persisted. This may be difficult to justify in light of recent evidence that polydrug, rather than monodrug, abuse has become an important pattern of usage, particularly in terms of emergency room visits (see Chapter 5). Confusion also occurs because substantive issues often become tainted with an emotionality that blurs objectivity. This is particularly important since funding for research, treatment, prevention, and training for both alcoholism and drug abuse often comes from the same limited budget.

Another factor that has obscured clarity in these areas is the lack of a theoretical framework for understanding the dynamic process of alcohol or drug dependency. For example, a moral model in which alcohol and other drugs are either good or bad; a judicial model in which they are legal or illegal; or a purely biomedical model in which alcoholism may be a disease but sedative-hypnotic abuse is not, are all of little assistance. Attempts to correlate specific personality types with a particular drug of abuse also has not proven very fruitful, since so many of the people who become dysfunctional because of excessive drug involvement often use more than one and frequently many substances.

In an attempt to develop a practical framework in which to view the many complex issues related to the problems of chemical dependency, Kissin[1] has described a biopsychosocial model of dependence. This model (see Figure I-1) assumes that there are biological, psychological, and social predisposing factors upon which the chemical agent acts. In an individual with a particular predisposition, chronic exposure to the drug will eventually produce dependence (psychological and/or physical). Once dependence does develop, continued use may result in a variety of medical, psychological, and social consequences, which now become superimposed upon and intermingled with the predisposing factors to create a particularly complex clinical picture. A common difficulty for clinical research in this area has been the problem of teasing apart findings that may have resulted from alcohol or other drug use as opposed to findings that may have predated and even predisposed to it.

However, as a point of departure in looking at the similarities and differences that exist between alcohol and other psychoactive drugs, this model should be helpful.

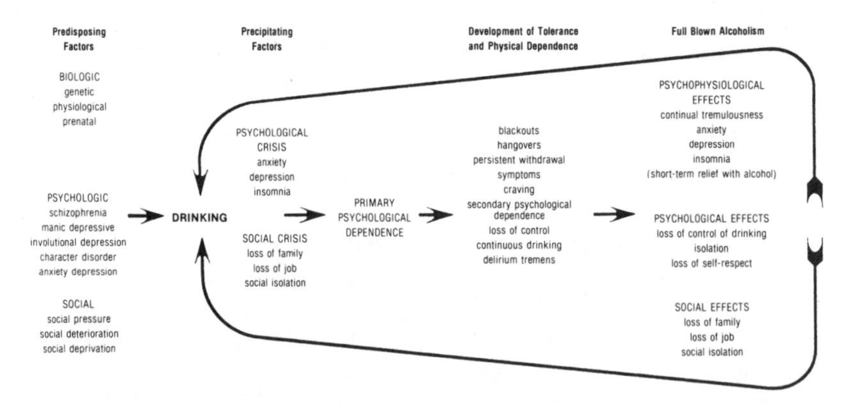

Figure I-1 Alcoholism as symptom and disease. Reproduced by permission from Kissin B: Theory and practice in the treatment of alcoholism, in Kissin B, Begleiter H (eds): *Treatment and Rehabilitation of the Chronic Alcoholic.* New York, Plenum Press, 1977.

Obviously, trying to fit all psychoactive compounds into every cell created by this model and then comparing them would be an awesome, redundant, and ultimately unproductive task. In deciding which areas deserved attention in this volume and which were either adequately covered elsewhere or were not germane to our objectives, certain subjective judgments had to be made. For example, there seemed to be little point in comparing alcohol to other drugs in the area of clinical pathology. This is covered very comprehensively in many places for both alcohol and other drugs, and a system-by-system comparison seems unnecessary here.

The chapters included in this volume were thought to present a variety of similarities and differences between alcohol and other drugs to warrant a systematic presentation. In structuring it this way, discrete disciplines may be examined uncontaminated by issues that are more appropriate to other disciplines. Through this format it is hoped that an objective and rational approach to the use, misuse, and abuse of alcohol and other psychoactive substances can be accomplished.

Joel Solomon, MD
Kim A. Keeley, MD, MSH

REFERENCE

1. Kissin B: Theory and practice in the treatment of alcoholism, in Kissin B, Begleiter H (eds): *Treatment and Rehabilitation of the Chronic Alcoholic.* New York, Plenum Press, 1977.

1 A Historical Review of Drug and Alcohol Use

Benjamin Kissin

The recent heroin epidemic in the United States during the late 1960s, as well as the amphetamine epidemics in Japan and Sweden since World War II, have brought into sharp focus the problem of what has come to be known as "drug addiction." However, from both social and medical points of view, these epidemics must be seen as merely occasional eruptions from a perpetual, underlying condition that affects many, if not all, societies. This state is the use, and sometimes abuse, of psychoactive, mind-altering chemical substances which persons at all times and in all places have used to help them adapt to the human condition. Accordingly, to obtain an overview of the role this practice has played in the evolution of human society, it is helpful to review historically what is known about the use and abuse of alcohol and other drugs.

ALCOHOL

A popular book on alcohol by Berton Roueché opens with the following paragraph:

> The basic needs of the human race, its members have long agreed, are food, clothing, and shelter. To that fundamental trinity most modern authorities would add, as equally compelling, security and love. There are, however, many other needs whose satisfaction, though somewhat less essential, can seldom be comfortably denied. One of these, and perhaps the most insistent, is an occasional release from the intolerable clutch of reality. All men throughout recorded history have known this tryanny of memory and mind and all have sought, and invariably found, some reliable means of briefly loosening its grip. The most conspicuous result of their search, if not the most effective, is a colorless liquid called ethyl hydroxide or, more popularly, alcohol. It is also the oldest, the most widely esteemed, and the most abysmally misunderstood.[1]

The history of drug dependence is the history of man's search for "the occasional release from the intolerable clutch of reality" through the taking into his body—by ingestion, inhalation, or injection—of some magical chemical substance. This need for a higher experience, which some have equated with the mystical aspect of human nature, has been the source of two of the most important of human institutions—religion and drugs. Inevitably, and particularly in the more primitive religions, the two are intricately related. But it is especially with mankind's experience with drugs that we are involved.

As with food and fire, man first discovered drugs in his environment. Alcohol was readily available wherever fermentation of plant starches occurred in the presence of yeast. Roueché[1] suggests as the most likely first sources of alcohol, fermented fruit juice (wine), fermented grain (beer), or fermented honey (mead). From the evidence of early language, he deduces that "mead" may have been the first, since the word "mead" or its derivatives in both the modern and the most ancient Indo-European languages stands for "sweetness" and "intoxication." In any event, we may infer from the evidence derived from the study of Stone Age cultures that have survived into recent times that early Neolithic man already had beer and wine. With the exception of three extremely restricted tribes (the Eskimos, the Australian aborigines, and the Tierra del Fuego natives), all known surviving, primitive societies have possessed some form of wine or beer. Roueché[1] goes on to speculate that early Neolithic agriculture was in some part directed toward the manufacture of beer, and quotes Edgar Anderson, who wrote, "Man may well have been a brewer before he was a baker." Roueché then goes on to state, "It seems, at any rate, significant that while alcohol has been found among pre-agricultural peoples, no instances are recorded of agricultural societies to which its existence was (or is) unknown."[1]

The common use of alcohol in the beginnings of history is well documented in the earliest writings of Mesopotamia and Egypt. Descriptions of drunkenness were frequent, as were prescribed remedies. The widespread use of alcoholic beverages was characteristic of all of the early civilizations, the Oriental, the Greek, and the Roman, although attitudes toward drunkenness varied widely from place to place.

The major change in the pattern of alcohol ingestion occurred about 800 AD when an Arabian alchemist named Geber developed the process of distillation. But Geber apparently saw no special value in distilled alcohol. As Roueché[1] points out, that discovery was made toward the end of the thirteenth century by Arnauld de Villeneuve, a professor of medicine at the University of Montpellier. Arnauld found that through repeated distillation of red or white wine, he could derive a clear, colorless liquid of great purity and power which, he wrote, "prolongs life, clears away ill-humours, revives the heart, and maintains youth."[1] On this noble fluid he bestowed the name aqua vitae. And with this generous title, distilled liquor was off to a long and inglorious career.

Alcohol, like all other drugs, can be and is used or abused, and its use and abuse throughout history is well documented. Ancient cuneiform and hieroglyphic inscriptions describe both the normal and abnormal use of alcohol. The Old Testament prescribes the use of wine in religious rituals but also alludes to Noah's excess: "and he drank of the wine and was drunken." Drunkenness as a sometime thing in certain individuals is suggested in the ancient writings, but the prototype of the drunkard is a more recent event. That development occurred in the seventeenth century through Franciscus Sylvius's discovery that aqua vitae could be obtained through the distillation of fermented grain (beer) as well as from wine, a discovery that permitted its production in vast quantities. The spread of distilled liquor as junever (the Dutch word for the flavoring herb juniper) in Holland, then as genièvre in France, gin in England, and vodka in Russia was accompanied by epidemics of drunkenness. Especially in England, where the government in 1690 in order to increase tax receipts passed an "Act for the Encouraging of the Distillation of Brandy and Spirits from Corn," the condition of alcoholism became so widespread among the poor as to threaten the existence of London lower-class society. By the middle of the eighteenth century, the situation had become so degenerate (as depicted in Hogarth's engravings) that parents and children alike in the slums of London were almost universally alcoholic. The government, finally recognizing the danger of social collapse, increased the taxes on liquor to a price beyond the reach of the poor, thus controlling the epidemic and simultaneously developing a steady source of tax revenues.[2] This doubly effective device, which acted both as a control of drunkenness and a source of revenue, has remained in all governmental practices ever since.

The "Great Experiment" in government control of alcoholism took place, of course, in the United States with the passage of the Eighteenth (Prohibition) Amendment in December of 1917. Despite the almost unanimous opinion that Prohibition was a failure, there seems to be no question that during its first few years in Canada, Finland, and the United States there was a marked drop in alcohol consumption and in alcohol-related problems.[3] Indeed, even though in the later years of Prohibition the rates for alcohol consumption and for alcohol-related problems rose slowly, at no time did they approach those that existed prior to or subsequent to the years when the law was in effect. Consequently, Prohibition was effective in accomplishing its original aim. What did contribute to its failure was the unforeseen development of social problems, particularly that of organized crime, which have remained a blight on our society to the present day.

Perhaps from our experience with Prohibition, we may gain wisdom in dealing with the specific problems addressed later in this volume. Legal intervention is a powerful instrument and laws often do accomplish their intended goals. However, sometimes, unanticipated ramifications and consequences that are not considered at the initiation of a law may ultimately invalidate or undermine its original purpose. Such was true with Prohibition; such may be true with the legalization or illegalization of marijuana, the combining of alcoholism and drug programs, or the universalization of methadone maintenance. This pattern seems to be particularly true where attempts are made to change strong patterns of social behavior through legislation.

CANNABIS (MARIJUANA)

Unlike alcohol, which could be and was produced wherever there was possible fermentation of various plant starches, most ancient psychoactive drugs were the products of specific plants. Hence the use and abuse of these drugs in the world was limited by the geographical frequency of their distribution. Among those plants with the widest world distribution is *Cannabis sativa,* also known as hemp, from which marijuana or its stronger version, hashish, derives. Hence it is not surprising that cannabis has been used in one form or another almost since the beginning of history and in all parts of the world.

The earliest references to the magical qualities of the hemp plant came from China, where a mythical emperor, Shen Nung, is alleged to have taught his people the value of cannabis as a medicine in the year 2737 BC. Shortly thereafter (and better documented) a Chinese physician, Hoa-Gho, mixed the cannabis resin with wine to produce a preparation called *ma-yo,* which was used as an anesthetic during surgery.[2] However, cannabis came to be used predominantly as a mind-altering substance in India and the Middle East, where references to its mystical properties are found frequently in the first millenium BC. The widespread use of cannabis as a mind-altering substance appeared to be more or less confined to India till about 1000 AD, at which time, in association with the Mohammedan conquests from Arabia into India and across North Africa, the use of hashish spread. The close association of cannabis use with the Mohammedan religion may be related to the specific prohibition by the Prophet of the use of alcohol. Certainly, hashish is most closely associated with the Arab Muslim faith. Specific references to hashish in the early Arab culture include that to Hasan-Ibn-Sabbah, the eleventh-century leader of the Assassins (a name presumed to stem from hashish, with which they were rewarded for their military exploits), and the frequent mention of bhang, called *beng* by the Arabs, in the medieval "Tales of Scheherazade."

Though the use of cannabis remained endemic in India, the Near East, and North Africa for over a thousand years, there appeared to be little tendency for its use to become widespread in Europe. The plant itself was well known to Europeans in the thirteenth century, and cloth made from its fibers were extensively used. Also, its mind-altering properties were well recognized, and François Rabelais described them fully in the sixteenth century. The presence of an alcohol-oriented society seems to have generally diminished interest in cannabis. This facet may be reflected in the fact that the ancient Greeks (an alcohol-oriented society) traded substantially with the people of the Near East, who freely used marijuana, yet they, the Greeks, never embraced the use of cannabis.

Not until Napoleon's soldiers brought back hashish from Egypt in the early nineteenth century was cannabis used to any extent in Europe. But even then it was taken up mainly by a few literati, such as Théophile Gautier, Charles Baudelaire, and Alexandre Dumas, who founded the famous "Le Club des Hachicheres" in Paris. Although these three wrote glowingly of the mind-altering effects of hashish, the use of the drug remained largely confined to a few cognoscenti and never really became popular among the people. This was at a time when alcoholism was rife in Europe, and again offers evidence that in the presence of alcohol, cannabis does not appear to flourish.

The history of marijuana in America is not unlike that in Europe.

The hemp plant was widely grown throughout the colonies and in the early United States as a fiber for clothing. The properties of the hemp extract (marijuana) were recognized and advocated for their medicinal uses. But, again, there was at first no widespread use of marijuana as a mind-altering substance. As Brecher puts it, "Not until the Eighteenth Amendment and the Volstead Act of 1920 raised the price of alcoholic beverages and made them less convenient to secure and inferior in quality, did a substantial commercial trade in marijuana for recreational use spring up."[2] The more recent widespread use of marijuana among young people in the United States appears to be the results, not of any scarcity of alcohol, but rather of a revolution against the values and attitudes of an alcohol-oriented society.

OPIUM

Opium is obtained from the opium poppy, *Papaver somniferum,* a plant that flourishes mainly in a warm, dry climate. Its present-day cultivation occurs mainly in a belt stretching from the Balkans through Turkey, Afghanistan, Pakistan, India, Indochina, and South China. Its first origins seem to have been in ancient Mesopotamia, where it was known 4000 years ago as "the flower of happiness." The Persians brought it to the Greeks who, in turn, lent it to the Romans. Later the Arabs improved on the extraction of opium from the juice of the unripe seed capsule of the poppy and traded it broadly, introducing the drug's use into India and China in the eighth century. By the fifteenth century its use was well established in India and China, and these countries began cultivating the opium poppy themselves.

However, before the seventeenth century, opium use in China and elsewhere was essentially medicinal, primarily for the relief of dysenteries, but also for the relief of pain and insomnia. Paracelsus (1493–1541) is credited with having first compounded the tincture of opium known as "laudanum" and of describing its appropriate therapeutic indications. The widespread use of opium as a mind-altering substance came about almost as an historical accident when the last of the Ming emperors, in the early seventeenth century, forbade the smoking of tobacco as a European vice. Opium smoking was then adopted as a substitute for tobacco and spread rapidly, particularly among the poor. The passivity associated with opium smoking as opposed to the more stimulant actions of nicotine resulted in a characteristic apathy among those afflicted.

When the Chinese authorities sought, through the curtailment of opium imports, to combat this sordid practice which threatened to undermine their society, the British retaliated with the notorious

Opium War (1839–1842), forcing the Chinese to import large quantities of opium from India. To become less dependent on English trade, the Chinese reestablished the cultivation of opium, and opium smoking became widespread throughout China and the Far East. When Chinese laborers were imported in large numbers to the West Coast of the United States in the 1850s, opium smoking was introduced into this country.

Meanwhile, in 1803, a German pharmacist, Serturner, had isolated the opium alkaloid morphine and fully described its medicinal properties, particularly its use as an analgesic. At the time of the American Civil War, the hypodermic needle and parenteral injection had only recently been invented. The huge number of battlefield casualties resulted in the widespread indiscriminate use of injected morphine and many survivors found themselves incurably addicted to morphinism, which became known as "the soldier's disease." The concomitant practices of morphine abuse by the soldiers and opium smoking by the Chinese immigrants introduced opiates into the national consciousness as an "ideal tranquilizer," and by the end of the nineteenth century their use was widespread among all strata of society in the United States.

It is generally estimated that in 1900 there were in the United States approximately 500,000 opiate abusers in a population of 100 million, or roughly twice the incidence in the American heroin epidemic of the 1960s. Opiates were freely sold in groceries and drug stores, without prescription, for every real and imaginary illness. The vast majority of addicted individuals were elderly middle-class white women who took gradually increasing doses of laudanum or some other popular opiate-containing nostrum. It is interesting that there was no strong anti-opium lobby as there was always a strong temperance movement. This was almost certainly the case because opium abuse, unlike alcohol abuse, usually resulted in marked passivity and produced no violent antisocial behavior. It is also of interest that the first anti-opiate legislation in the United States was passed in California in 1875 and was directed at keeping whites from associating with opium-smoking Chinese and also toward increasing the working capacity of the Chinese. The concern of the authorities about opium smoking gradually spilled over onto the less socially disreputable practices of opium and morphine ingestion, and ultimately resulted in the restrictive legislations of the Pure Food and Drug Act of 1906 and the Harrison Narcotic Act of 1914.

The next episode in the opiate drama began in 1898 with the synthesis of diacetylmorphine, better known under the name of *heroin.* This drug was originally touted as being a stronger analgesic than morphine yet less addictive! Only the first part of this statement proved to be correct. Heroin was used early in the nineteenth century as a powerful anti-cough mixture and only rarely as a euphoriant drug. Gradually, however, it began to replace parenterally injected morphine as the drug of choice for the euphoriant "high." This practice remained largely endemic in the ghettos of the cities and among "artistic" groups till the late 1950s and early 1960s, when it broke out into the virulent epidemic of the latter years.

Finally, methadone was synthesized in Germany during World War II as a synthetic narcotic used to substitute for the unavailable morphine. Its use in this country was first developed by Isbell and his group in the early 1950s at the Lexington, Kentucky, United States Public Health Service Hospital for addicts as the approved technique for detoxification from opiate addiction. Dole and Nyswander in 1964 modified the use of methadone into the concept of methadone maintenance, which today represents the major chemotherapeutic approach to heroin addiction. However, the saga of opiate addiction is not yet finished, since illicit methadone now constitutes a major source of opiates on the streets of New York City.

CAFFEINE AND NICOTINE

These drugs, which form the active ingredients of tea and coffee and then tobacco, respectively, have become so very much a part of our normal everyday existence as not to be considered drugs at all. Yet tea is, next to water, the most commonly consumed liquid in the world and coffee is not far behind. In terms of the worldwide use of psychoactive drugs, caffeine is by far the most frequently used, with nicotine, alcohol, and cannabis following in that order. Despite the mildness of the stimulant effects of caffeine and nicotine, there is evidence of at least moderate dependence on these drugs. Consequently, any in-depth consideration of the drug scene must certainly deal with these two, the most commonly used drugs of all.

Caffeine

Tea is a beverage made from the leaves of a bush, *Camellia (Thea) japonica,* indigenous to East Asia. Although its use was probably widespread in China earlier, tea is first mentioned in Chinese literature in the third century AD and by the eighth century it was being cultivated commercially.[2] It was introduced into Europe by the Dutch East India Company around 1600 and became particularly popular in England. It was the leading beverage in the American colonies until after the Boston Tea Party, after which it was superseded by coffee.[2] Tea is now grown mainly in the Far East in areas with a rich, light soil; a warm, humid climate; and a moderate rainfall. It is exported to most of the countries in the world.

The coffee shrub, *Coffea arabica,* is believed to have originated in Ethiopia and to have been introduced into Arabia in the fifteenth century AD. The first historical reference comes from Ethiopia about 1000 AD, at which time it was first used as a food. Coffee as we know it was first made from the roasted, ground beans after its introduction to Arabia, and its use spread rapidly. Because of the prohibition on alcohol, coffee soon became the universal beverage of Mohammedanism. It subsequently spread to Europe in the mid-seventeenth century, where it became associated with the development of coffeehouses, the forerunner of the modern club. Coffee is now grown commercially in hot, humid, tropical areas, mainly equatorial Africa and Brazil, and exported throughout the world.

The effects of caffeine in tea and coffee are those of a mild stimulant to which people everywhere rapidly become habituated. Other beverages in which caffeine is found are the cola drinks made from the kola nut, which comes from West Africa; hot chocolate from the cocoa tree of Central and South America; and maté, a Paraguayan tea made from the ilex plant found in South America.

Nicotine

Tobacco was indigenous to the Americas and hence unknown to the civilized world until after Columbus's discovery. The spread of its use in Europe after its first introduction in the early sixteenth century was almost epidemic. At first, because of its scarcity and high price, it was a malady only of the rich and in 1610 an English observer wrote, "Many a young nobleman's estate is altogether spent and scattered to nothing in smoke." By 1614 the number of tobacco shops in London had risen to 7000 and smoking had become a vice of the poor as well.

Popes Urban VIII and Innocent X issued papal bulls against tobacco in 1642 and 1650, respectively, but without effect. The Sultan Murad IV decreed the death penalty for smoking tobacco, but nevertheless the continued spread of smoking persisted and became almost universal throughout the world. As Brecher puts it, "From those days until today it is important to note, no country that has ever learned to use tobacco has given up the practice...More remarkable still, no other substance has been found through the centuries since 1492 that can take the place of tobacco. Tobacco smokers who learn to smoke opium or marijuana go right on smoking tobacco in addition—clear evidence surely that it is something in the tobacco rather than the act of smoking which underlies the addiction."[2]

This something is presumably nicotine and, again quoting Brecher,[2] "Nicotine produces a unique combination of effects: at moments when stimulation is needed, smokers perceive the smoke as stimulative and when they feel anxious, they perceive the smoke as tranquilizing." This dual effect of nicotine is similar to that of alcohol, which can be both stimulant or relaxant, depending on the dose and the underlying mood, and may account for the universal popularity of both of these substances. It may also account in part for the fact that alcoholics are notoriously heavy smokers, while ex-alcoholics remain heavy smokers but become vast consumers of coffee and cola drinks.

LESS COMMON BOTANICAL PSYCHOACTIVE SUBSTANCES

This brief historical review of drug use reveals several patterns that are instructive in considering why certain substances rapidly receive widespread and universal acceptance and why others do not. Despite our emphasis on the historical significance of availability, it is important to observe that availability per se, although a major determinant of drug use, is not the only one. Certainly, where no drug is available, no use is possible; and where the drug is naturally abundant, as with alcohol and cannabis, its free use may develop. On the other hand, neither coffee nor tobacco was introduced into European culture until relatively recently, and they have become the most widely utilized drugs of all. Obviously, in the historical development of drug use, other factors have played major roles, some leading to increasing spread, other restricting it. Historically, in addition to availability, some factors that appear to encourage the widespread use of a drug are 1) the psychological satisfaction to be derived therefrom, 2) a relative mildness of effect (caffeine and nicotine are more widely used than heroin and LSD), 3) a broad spectrum of pharmacologic effects, both dose related and mood related (as with nicotine, alcohol, and cannabis), and 4) a deliberate campaign by a major social agency to increase drug use for economic gain (the British Corn Laws and alcohol, the British Opium Wars and opium, and the American underworld and heroin). These considerations are important in discussing the present group of "less common botanical psychoactive substances" and in explaining why they have received relatively less widespread acceptance than the drugs described thus far.

Cocaine

Cocaine derives from the coca leaf of the plant, *Erythroxylon coca,* which is indigenous to South America and particularly to the mountainous areas of Peru and Bolivia. The use of the coca leaf as a wad combined with lime (to liberate the active alkaloid) held in the inner cheek, is documented in pre-Columbian ceramics of Peru (Mochica) and Ecuador (Harachi) dating back to 500 AD. This custom has continued up to the present, particularly among the mountain Indians of Peru and Bolivia, who use coca leaves to combat fatigue and the depression of poverty. There is no evidence that this use of the drug is "addictive" since these people readily abandon the habit when transferred to the coast.

The use of coca leaves with the concomitant stimulation and occasional euphoria was never adopted by the Spaniards nor brought back to Europe to any great extent. The reasons for the failure of the chewing of coca leaves to become a widespread practice are not readily apparent. In terms of the four characteristics listed earlier, the drug qualifies eminently to assume a distribution of epidemic proportions. It is relatively harmless,* it provides one of the most euphoriant of psychological effects, and it does have a broad spectrum of pharmacologic effects, from mild fatigue relief in small doses to manifestations imitative of a psychosis with large doses of cocaine. However, it is unclear whether the psychotomimetic effects of large doses are ever obtainable through the limited route of coca-leaf chewing alone or whether they can be obtained only with parental cocaine. Perhaps its use has never been greater because cocaine is destroyed by the intestinal juices and by smoking, so that it can be absorbed only through mucosal tissues or by injection. It may be that chewing coca leaves, like chewing tobacco, is considered socially unacceptable.

In any event, for whatever reasons, the widespread use of the coca leaf never developed until the middle of the nineteenth century. At that time, two simultaneous developments occurred that defined the present status of cocaine. The active ingredient of coca leaves, the alkaloid cocaine, was isolated in 1844. It received little notice until 1883, when a German army physician, Theodor Aschenbrandt, found it effective in relieving fatigue among soldiers. Sigmund Freud published a glowing report on its salutary effects in 1884 and recommended its use for a variety of medical and psychological ailments (including its use as a local anesthetic). However, shortly thereafter it

*Brecher quotes Richard Schultes, Director of the Botanical Museum of Harvard as having chewed coca leaves daily during his eight years of exploration in the Amazon Valley without becoming "addicted" or suffering any apparent physical harm.

became apparent that the excessive use of cocaine led both to "addiction" and to a paranoid psychotic state which was often irreversible, and the use of cocaine as a psychoactive drug rapidly fell into disrepute.

At about the same time, in 1886, a dispenser of patent medicines in Atlanta, Georgia, named John Styth Pemberton, developed a syrup concocted from a mixture of the coca leaf and an extract of the cola nut—the African product containing caffeine. He named the resultant drink Coca-Cola. This drink soon assumed an enormous and worldwide popularity. However, it is unclear how much, if any, cocaine effect was realized since, as previously mentioned, the drug is destroyed by the intestinal juices when ingested orally. In any event, after the passage of the Pure Food and Drug Act of 1906, the coca leaf extract was removed from Coca-Cola, leaving only a mild caffeine content.

Plant Hallucinogens

A variety of hallucinogenic chemicals have been extracted from various plants that are indigenous to the Western Hemisphere and which have been used by different Indian tribes for as long as 3500 years. Among these chemicals are mescaline, psilocybin, DMT, and substances closely related to LSD. The most widespread use of hallucinogens by the American Indian was that of peyote, a spineless cactus with a small crown or "button," which, when sliced off and dried, formed a disk known as the mescal button. This button, when swallowed, produced the characteristic hallucinogenic experience involved as part of the religious ritual of the ancient Aztecs and the Mescalero* Apaches. The peyote ritual is still legally practiced by Indians throughout the western United States.

The Aztecs also used a variety of mushrooms to achieve hallucinogenic effects during ceremonial rituals. Several of these species have recently been identified as still in use in parts of Mexico. The active ingredients are psilocybin and psilocin. The Aztecs also employed hallucinogenic morning glory seeds, which recent research has found to contain a substance related to LSD. Still another hallucinogen, yurenia, was and still is used by primitive Indians of eastern Brazil. The active ingredient has been identified as N,N-dimethyltryptamine, better known as DMT.

Further afield, hallucinogenic mushrooms known as fly agaric have been known to be used by Siberian tribesmen since the eigh-

*From which the name of the active ingredient, mescaline, stems.

teenth century. Nutmeg also has been used for its hallucinogenic properties both among the American Indians and in Southeast Asia.

SYNTHETIC PSYCHOACTIVE SUBSTANCES

The great age of synthetic organic chemistry began in Germany in the early nineteenth century. Perhaps the first psychoactive drug to be synthesized was chloral hydrate, discovered by Liebig in 1832. In 1869 Leibreich predicted, on the basis of the chemical relationship of chloral hydrate to chloroform, that it might have significant hypnotic properties. This proved to be true and to this day the drug remains one of the most valuable and effective sedatives.

The synthesis of psychoactive drugs reached a new high point when von Baeyer synthesized barbituric acid in 1864. The original breakthrough was followed by the synthesis of over 2500 barbiturates, of which over 50 have been marketed for medicinal use. The first effective barbiturate hypnotic was barbital, introduced into medicine by Fischer and von Mering in 1903 under the name of Veronal. The second oldest significant barbiturate is phenobarbital, introduced in 1912 and marketed under the trade name of Luminal. The barbiturates rapidly replaced almost all other sedative and hypnotic agents in medical use so that by the late 1930s an estimated one billion grains were being prescribed each year in the United States alone.

Barbiturates were used mainly as hypnotics and sedatives but also as anti-anxiety agents, although the value of the drugs was limited in this last therapeutic category because of the sedative side effects. In 1952 Delay's discovery that chlorpromazine had marked anti-anxiety as well as antipsychotic therapeutic efficacy introduced the new concept of the "tranquilizer." This concept was further extended with the introduction by Berger in 1954 of meprobamate (Miltown), which provided mild anti-anxiety effects with minimal drowsiness. This class of agents was further extended in 1961 with the discovery of chlordiazepoxide (Librium), which soon became the most widely prescribed tranquilizer of our times.

On the other side of the "affective" coin, the first major stimulant and appetite suppressor, introduced in 1935, was amphetamine. This was used to combat mild fatigue, to increase energy, to allay appetite and, in some instances, to treat depression. However, a specific antidepressant regimen was not described until 1952, when Delay reported that the monoamine oxidase inhibitors that were being used in the treatment of tuberculosis were also effective in treating depression. The MAO inhibitors were largely replaced in 1958 by the tricyclic antidepressants, which were found to be equally effective but far less dangerous than their predecessors.

Intercurrently Hofman, in 1943, discovered LSD and vividly described its hallucinogenic and psychotomimetic properties. The popularization of these effects, combined with a disenchanted youth after World War II, initiated an interest in "psychedelic" experiences and renewed involvement with the plant hallucinogens, mescaline, psilocybin, and cannabis. These developments led, in the late 1950s and early 1960s, to the unfolding of a drug culture so powerful that it influenced the drug-taking pattern of the world.

The commercial success of the synthetic hypnotics, minor tranquilizers, stimulants, and antidepressants encouraged the drug companies to develop even more modifications and a spate of psychoactive drugs flooded the market in the 1960s. Among these were the sedatives glutethimide (Doriden), methyprylon (Noludar), ethchlorvynol (Placidyl), and methaqualone (Quaalude); the minor tranquilizers diazepam (Valium) and oxazepam (Serax); new stimulants methamphetamine and phenmetrazine (Preludin); and a variety of tricyclic antidepressants. Each of these drugs became fashionable for a time for their prescribed medical use and then, in turn, became popular as "fads" in the drug culture. The vast number of prescribed and illicitly available synthetic drugs appeared to contribute to the drug culture in the early 1960s, which led naturally to the heroin epidemic of the late 1960s and to the polydrug and multidrug abuse patterns of the 1970s.

REFERENCES

1. Roueché B: *The Neutral Spirit.* Boston, Little, Brown & Co, 1960.
2. Brecher M: *Licit and Illicit Drugs.* Boston, Little, Brown & Co, 1972.
3. Popham RE, Schmidt W, deLint J: The effects of legal restraint on drinking, in Kissin B, Begleiter H (eds): *The Biology of Alcoholism: vol 4. Social Aspects of Alcoholism.* New York, Plenum Press, 1976, pp 579–626.

2 Sociocultural Aspects of Alcohol and Drug Use and Abuse

Joseph Westermeyer

OVERVIEW OF CULTURE AND INTOXICANT USE

Prior even to written history, people in virtually all societies used psychoactive compounds. Diversity in compounds and modes of administration have rivaled our species' accomplishments in agriculture, transport, and other areas of human endeavor. Aboriginal North Americans ignited tobacco and other plant substances to inhale them into their lungs, while South Americans pulverized hallucinogenic substances in order to eat them, blow them as a snuff into their nasal passages,[1] or take them as enemas.[2] Eurasians ingested cannabis and the resin of the opium poppy. Africans discovered several stimulant-type compounds (eg, chat, kola, yohimbine).[3-5] Andean mountaineers chewed coca leaf as a stimulant,[6,7] and Australo-Asians took betel by the same method and for the same purpose.[8] Herders and farmers throughout the world found that alcohol could be obtained from virtually any carbohydrate source—from horse's milk to certain cacti to various grains and fruits and tubers.

Intoxicants widely used within a society became integrated within the norms and values of the society.[9-11] Religious functions generally accompanied intoxication among tribal peoples, such as the use of stimulants and hallucinogens for the vision quest in the New World and ceremonial use of alcohol in the Old World. Social controls were exerted over intoxication, with prescribed and proscribed times for use. Often there was a social imperative that a person use a psychoactive compound, usually alcohol or hallucinogens, in particular social contexts whether or not the individual chose to use it. Notable exceptions to this social imperative have been opium and certain stimulants.[12]

Most societies seem to require many centuries to weave a psychoactive substance into their cultural warp and woof.[13,14] Northern and western Europeans have had alcoholic beverages for a thousand years or more, but still exert uneven social control over it—albeit more effectively than hundreds of years ago.[15,16] American Indian groups have had alcoholic beverages for 200–400 years,[17] yet many still encounter difficulty with its use.[18] Rapid expansion of drug use of all types among the world's youth over the last decade has again demonstrated the vulnerability of populations to the introduction of new drugs not part of a society's norms and mores.[19-21] Rapid sociocultural changes can also influence even long-entrenched patterns of intoxicant use, as exemplified by Japanese drinking over the last few decades.[22]

Ethnocentrism—the relatively high valuation most of us ascribe to our own society, with lower valuation to other societies—makes it difficult to assess objectively the interrelationship of intoxicants and society. As a case in point, most Africans, Europeans, and Americans view alcoholic persons as more conformist than opiate addicts. And that view has some sociological validity in those regions, at least as reflected in employment and marriage rates. But in parts of Asia the opium addict more likely is married and employed, while the alcoholic individual more likely is unemployed and single or divorced, ie, more socially "deviant."[12,23]

Intoxicants—like sex, money, and power—lend themselves eminently to individual and social symbolism. Wine can be a sign of Christ's blood to Roman Catholics, or a sign of the devil to some American Indian cults and certain fundamentalist Protestant groups. Drinking has been used as an instant status symbol by youth, as well as by colonized people seeking to emulate their colonizers.[24] It has signaled changing political relationships between ethnic groups,[25] and it has contributed to the maintenance of ethnic boundaries between adjacent cultural groups.[26] Cannabis has represented friendship and hospitality among some Moslem groups, and an aphrodisiac[27] or antiauthoritarian symbol among contemporary youth. Morphine may bring welcome relief to a surgical patient in Manhattan, while its chemical cousin, heroin, may represent all that is evil and dangerous in Manhattan to that same surgical patient.

Societies have traditionally governed intoxicant use primarily by religious stricture. Islamic groups stringently forbid alcohol use, but frequently ignore widespread cannabis use.[28] Membership in the Native American Church requires peyote intoxication, but alcohol drinking is frowned upon.[29] Mormonism opposes any personal or recreational use of psychoactive compounds, including tobacco and caffeine. While religious proscription has been an effective means for governing use of intoxicants in many populations, the move towards secularism and away from organized religion in literate societies has impeded this means for moderating or preventing drug and alcohol use in numerous populations today. Many societies undergoing rapid sociocultural change have experienced more frequent and less controlled use of recreational intoxicants.[30]

Increasingly over the last several decades societies have begun to use law and law enforcement as a means of regulating intoxicants. Some such efforts have met with relative success, as in the communist nations of eastern Europe and Asia where illicit opiate addiction is virtually nonexistent.[31] Other societies have had some success in reducing intoxicant abuse but have paid a high social cost in corruption of police and officials, eg, alcohol prohibition in the United States during the 1920s. Opium prohibition has often led to decreased use of opium in several countries, but increased use of heroin.[32,33] Among nations with less coercive political systems, control of substance abuse presents a major and growing social challenge.

SOCIAL FUNCTIONS OF DRUGS AND ALCOHOL

Religion

Much of what we deed to the Supreme Being can also be ascribed to the community or society in which we live. It is our community that gives us our life, nurtures it, and can take it away. Our society is the immortal, the omnipotent, the omniscient—at least in relation to our own individual mortal, vulnerable, and limited perspective. We have faith in our society, hope in it, trust it, love it, sometimes even die for it—as people have done for their God or gods over the eons. Or we can hate our society as we can hate the Supreme Being—a condition described by anthropologists as anomie and by religionists as despair.

Society and religion epitomize the conflict and the romance, the dependence and the ambiguity that mark relations between each individual and society at large.

It should come as no surprise, then, that communal intoxicant use often accompanies religious experience. Wine as the "blood of Christ" in Roman and Orthodox Catholic services, and as an integral part of Judaic rituals, has its roots in the ancient religions of Asia Minor.[34] Various alcoholic beverages play similar communal-religious roles in cultures around the world, from the Papago of southwestern United States[35] to an Afro-Brazilian cult[36] to a Gond village in India.[37] Hallucinogens and tobacco have served such purposes in North and South America[38,39]; the hallucinogen peyote remains the sacramental intoxicant of the pan-Indian Native American Church.[29]

Parenthetically, religious practice can provide a path away from chronic, problematic intoxication. Fundamentalist religion has provided an alternative to some chemically dependent young Americans in the last decade.[40,41] Zapotec-mestizo alcoholics in Mexico and Cambian alcoholics in Bolivia have converted to abstinent Protestant sects as a method for ameliorating drinking problems.[42,43] A Buddhist temple in Thailand has been successful in aiding some opium addicts to abandon the pipe.[44] Typically, such conversions involve not only a theological reorientation, but also changes in companions, in day-to-day behavior, and in values and attitudes.

Time Out

All societies provide temporary "time out" from usual social roles, work, everyday responsibilities, and social strictures. Examples in the United States are July 4, Labor Day, Thanksgiving, and New Year. Some are specific to certain social roles, eg, Mother's Day, Father's Day, Secretary's Day. Religious, ethnic, and clan festivities supplement national holidays. Individuals celebrate their own *rites de passage* (birthday, anniversary of marriage). These provide respite, a change of pace, relief from boredom, and a counterpoint to stability and effort.

Altered states of consciousness are a common feature of social "time out." Sometimes this is accomplished without intoxicants by fast, ecstatic dance, or prolonged chant (eg, aboriginal tribal groups of North America), risk-taking (eg, "rattlesnake cults" in the Appalachians), or silence (eg, some Quaker groups). Various intoxicants are typically employed for the same purposes, as exemplified by drinking rituals in Asian, European, and American celebrations.

Social Coping

Intoxicants are not used socially just on ceremonial or ritual occasions. They are also used to facilitate secular behavior. Work often constitutes such a raison d'être. Stimulants have been most widely taken for this purpose: coffee and tea breaks in the Americas and Europe, betel chewing in Asia and Oceania, chat in the Middle East, kola in Africa, and coca leaf in the Andes. Alcohol has been taken before prolonged, repetitive work: during heavy labor or farm work, as among Peruvian peasants[45,46]; and in the assembly-line milieu required by mass production, as among British factory workers early in this century.[47] Drinking has accompanied bargaining in such widely diverse settings as business-lunch cocktails among executives in the United States and barter between tribal chiefs and merchants in Asia.[12] Opium has been used during work by peasant farmers and laborers in Asia.[12,48]

Improved sexual function—especially the relief of premature ejaculation in the male and anorgasmia in the female—has been given as the motivation for using opium, heroin,[49] amyl nitrate,[50] marijuana,[51] alcohol,[52] and amphetamines.[53] Certain social functions such as cocktail parties, Indian pow-wows, and Buddhist festivals commonly engender anxiety, which is often relieved and social interaction thereby enhanced by intoxicant drugs.

SOCIAL EFFECTS OF ALCOHOL AND DRUG ABUSE: SIMILARITIES

Work

If substance abuse begins during childhood or adolescence, it can interfere with the acquisition of work-related habits and skills. People affected early may not complete school or vocational training. Perhaps even more importantly, they often do not develop habits and attitudes of industriousness, persistence, and responsibility. Their time frame for gratification is within the next few hours, eg, the next drink of alcohol or heroin "hit," not within the next few weeks, eg, the next paycheck, or even the next few weeks, eg, completing a course of training or planning for promotion. Rehabilitation of such socially disabled individuals later in their twenties or thirties is especially difficult.

Those who encounter onset of substance abuse in their twenties or later in life usually have training, an occupation, and positive work habits and attitudes. Nonetheless, substance abuse affects their relationships with other workers, their efficiency, and their reliability. They shift their burdens to co-workers, or become irascible in their roles as a supervisor, co-worker, or supervisee. Their work becomes slipshod, and they are prone to on-the-job injury. Monday morning absenteeism and frequent sick leave ensue due to binges or to increased illness and injury.

As a result of increasing expenditures on alcohol or drugs, the substance-abusing person may take a second job. This reduces the commitment and energy available for the primary job. Theft from the workplace may supplement income to purchase alcohol or drugs. As work-related problems mount, the person is fired. Subsequent jobs often involve less income and/or less occupational prestige, and are held for progressively shorter periods of time.

Boring or repetitive work has been implicated by some as a cause for substance abuse, but there are few data to support this notion. For example, a survey among trade union members in an automobile industry indicated that their rate of alcoholism was less than that of the general population.[54] On the other hand, physicians have been noted to have relatively high rates of substance abuse although their work is the antithesis of mass production.[55,56]

Finances

Financial problems that accompany substance abuse have three origins. First is the cost of alcohol and the drug itself. Alcoholics commonly spend $2000–$3000 per year on alcohol, although it may be as low as several hundred dollars per year or as high as several thousand, depending on the type of alcoholic beverage, mode of purchase—liquor store *vs* cocktail lounge—and amount drunk. Smoking cannabis several times daily can readily cost similar amounts, as can heavy use of hallucinogens or stimulants. Heroin addiction typically runs much higher, from several thousand dollars per year to several tens of thousands. Prescribed drugs tend to cost least, even in addictive doses: a few hundred to several hundred dollars per year.

Second, substance abusers have increasing difficulty in obtaining money as their job efficiency decreases. Even if they obtain money illicitly, they become increasingly apt to make mistakes and to be caught. Chemically dependent thieves and burglars are no more successful at their chosen occupation than are legally employed workers with similar problems.

Third, many indirect costs are associated with substance abuse. Two major financial drains are medical care and legal fees. Illness and injuries related to substance abuse—such as car accidents, attorney's fees, court costs, increased vehicular insurance, fines, and days lost from work—become a heavy fiscal burden to the substance abuser. Like Scrooge's chain, it tends to grow of itself, with each financial problem tending to precipitate others. Bankruptcy frequently ensues in later stages.

Sexuality

Although intoxicants may enhance sexuality during early use, the effects of chronic substance abuse on sexuality are almost universally detrimental. Alcoholics and drug addicts report decreased libido, decreased frequency of intercourse, and increased sexual dysfunction. With rehabilitation, an unanticipated event for the recovering substance abuser is increased libido, which may further complicate a troubled marital relationship.[57]

Marriage and Family

As the person's financial crunch becomes worse, this has its effects on the family. A drug-abusing teenager may steal from other family members, or sell the belongings of the family, to obtain drugs. A housewife may use grocery and clothes money to purchase alcohol. If the father has been the sole wage earner, the wife and children may be forced to work in order to support themselves.

Impaired sexuality between the marital pair can have consequences for the marriage. A substance-abusing person, impotent or anorgasmic with his or her spouse, may seek an extramarital partner. Or the non-abusing spouse, anorgasmic or impotent due to anger with the substance-abusing spouse, may seek a sexual liaison outside the marriage. Or either partner may simply simmer angrily, not seeking an outside partner but becoming more sexually frustrated.

Most destructive to the family are the psychological, behavioral, and relationship concomitants of substance abuse. The substance-abusing family member creates crisis after crisis, is emotionally unstable, relates to others now as a child and then as a tyrant. Increasing this undermines the trust, concern, and mutual dependence of family members on each other. Children in such households frequently are underachievers at school and display a variety of behavioral problems (eg, school phobia, runaway, truancy, delinquency). A large minority of substance abusers come from such backgrounds themselves.[58]

or tickets

Darren

Dad

Social Affiliation

As substance abuse becomes a central theme in the person's life, there is increasing alienation from kinship group, family, friends, neighbors, co-workers, and others (eg, sports buddies, fellow church members, bridge group). Less time is spent with these groups, and with former hobbies and interests. Increasing amounts of time are spent seeking, using, and experiencing the intoxicant. At the same time that the substance abuser chooses to spend time apart from social peers, society likewise tends to isolate and blame the person who cannot abstain from or use drugs in a problem-free fashion.[59]

Some substance abusers join a subculture. These include many skid row drinkers,[60,61] young "street" drug users,[62] heroin users,[63-65] tavern or cocktail lounge alcoholics,[66] and even some Asian opium addicts.[67] They form loose-knit affiliations, have subgroup norms and values, meet at specific times and places, and often have their own jargon or dress. It seems likely that the intoxicant experience reinforces social affiliation in such groups, and that the group experience in turn supports the substance abuse.[68,69] Other substance abusers remain isolated. The latter group typically includes schizoid individuals, housewives, and professional people.

Health

Regardless of type of substance, chronic intoxicant use has certain effects on health. As one might expect from the psychomotor effects of intoxicants, trauma is a common sequela. Falls, burns, vehicular accidents, and fights account for most trauma cases. Infections are frequent; these particularly affect the respiratory system, the urogenital system, and the skin. Malnutrition occurs in association with decreased appetite and poor nutritional balance, as more money is spent on drugs than on food. Dental problems are a complication of poor oral hygiene, trauma, infection, and malnutrition. Mortality is also more frequent among substance abusers than in the general population. Accidents, suicide, homicide, and pneumonia account for many deaths.[70]

Legal Problems

Substance abusers appear in attorneys' offices, courts, jails, and prisons more frequently than do their fellow citizens. Divorce, bankruptcy, driving while intoxicated, vehicular accidents, manslaughter, assault, homicide, theft, child neglect and abuse, and forgery occur to a greater extent in this group.[64,71–73]

SOCIAL EFFECTS OF ALCOHOL AND DRUG ABUSE: DIFFERENCES

Rapidity of Course

Certain drugs lead to onset of behavioral, social, and health problems more rapidly than others. For example, on the average, alcoholic Americans present themselves for treatment a few to several years later than do American heroin addicts. Individual differences do exist, so that a few alcoholics encounter problems earlier than heroin addicts, but the mean differences vary widely. Even within the opiate drugs, heroin addiction in Asia leads to problems motivating treatment years earlier than does opium addiction.[74,75] Despite the difference in rates at which problems develop, the types of problems that ensue greatly resemble each other regardless of whether the drug is alcohol, opium, or heroin.

In addition to pharmacologic and individual differences, the social context of addiction can play a role. Expatriate Americans using opiates in Asia become addicted more rapidly than addicts in the United States. Expatriates also present for treatment sooner than "at-home" addicts.[76] This finding suggests that, in the absence of family and community relationships, addiction runs a more rapid course.

Cost

As previously noted, various types of alcohol and drugs vary widely in cost. An alcoholic, imbibing the same amount of alcohol in a year's time, can spend $1000 a year on inexpensive, fortified wine or $10,000 a year drinking fine scotch and cognac in a cocktail lounge. Sedative or opiate abusers forging prescriptions and tricking physicians into prescribing for them can remain intoxicated for only a few hundred dollars per year. On the other hand, heroin addicts in the United States may spend as much as $200 per day on their habit.

Just as drugs differ, individuals also differ. Some heroin users spend less per day on their drug than some alcoholics, but they are an exception. Individuals may also vary over time, using more drug or alcohol during some periods and less or none during other periods. Times change as well; there have been times in the United States when an opiate "high" was as cheap as an alcohol "high," and it still is in India according to Chopra.[48]

Legal Issues

Legal differences between alcohol and drugs fall into two categories. First is the legality of the substance itself. Alcohol, caffeine, and nicotine can be purchased in the United States and used legally as psychoactive substances. That this is true does not make it immutable; these same substances have been illicit in various times and places. Even tea was once widely opposed in Europe and North America, as noted by Greden.[77] But their legal status here and now does not place the user in legal jeopardy. Use of illicit substances—heroin, cannabis, certain hallucinogens—does put the user at civil risk. Further complicating matters is the medical use of drugs; certain drugs can be used licitly if prescribed by a physician, but are illicit if used for recreational self-intoxication or self-treatment (eg, morphine, cocaine, barbiturates). Society inevitably must choose which drugs are condoned as intoxicants and which are not, and history tells us that these decisions are prone to change over time—now being more lax, then more strict, and vice versa.

The second issue, the legal status of specific drugs, is often an urgent one from the perspective of society. This is due to the inability of some substance abusers to pay for their intoxicant substance, whether it be a licit or illicit substance. Faced with this problem, many substance abusers who have never stolen anything will steal the substance, or steal money to purchase the substance, or steal goods to exchange for money. (Of course, some substance abusers would oppose stealing under any circumstances, but their resolve may dissolve during intoxication.) Rather than steal, some choose other illicit means for obtaining funds, such as prostitution or forgery. Some even seek treatment rather than break the law. To be sure, not all felons with a substance-abuse problem break the law merely due to their pursuit of self-intoxication; many broke the laws before abusing substances and continue to break them when not abusing substances. And even though drug abuse in the United States is often associated with criminality, some data suggest that alcoholism accounts for more crime than does drug abuse.[73]

This drug legality issue is particularly important as a sociopolitical matter. As substance abuse progresses, the individual seeks larger doses and more frequent use and thus needs more money for drugs. At the same time, the ability to generate money becomes impaired due to work inefficiency. And the availability of money for alcohol or drugs may decrease as medical and legal fees rise. Faced with these circumstances, the substance abuser can be strongly motivated to step outside religious, social, and political strictures.

Note that it is not only drugs per se that accounts for the social consequences just described, but also society's laws regarding them. In parts of Asia, traditional opium smoking has not been associated with crime.[12] In contrast, opiate addiction in the United States has typically been associated with criminality.[78]

Mode and Frequency of Administration

Substances vary in how they may be taken. Opium can be eaten, smoked, or drunk in various infusions, but not injected. Another opiate, heroin, can be smoked or injected, but is not highly active if ingested. Modes of administration affect the kinds of medical problems that result.

For example, oral ingestion of opium is not associated with any particular disease.[23] Appearance of opium smoking in Asia, after this method was "imported" from the New World, led to widespread opium addiction during the eighteenth century, and the subsequent appearance of chronic pulmonary disorders.[79] Subsequent morphine and heroin extraction from opium, and the development of injection as a means of drug administration, have led to a whole new spectrum of addiction-related diseases in the nineteenth and twentieth centuries. These include anaphylactoid shock, bacterial endocarditis, osteomyelitis, phlebitis, angiitis, and myriad other problems. Even episodic, non-addictive use of heroin by injection can be fatal if the apparatus is dirty or the dosage unexpectedly high. From a social perspective, these increased morbidity and mortality rates place a heavy burden on the populace in terms of medical care, disability, and death of people during their productive years.

Frequency of administration is an important consideration. Episodic use of alcohol or tobacco in small or moderate doses rarely produces problems, but repeated doses of alcohol or tobacco throughout the day, over a period of years can and does predictably lead to untoward consequences for many people. This has social consequences in that a society may want to approve episodic use of some substances under some circumstances, while condemning frequent use of the same substance. Due to the risk involved, some substances or modes of administration may pose too great a risk for a society to condone, such as LSD or phencyclidine (PCP), where psychosis may be a complication.

Demographic and Lifestyle Differences

While the core substance abuse phenomenon remains remarkably constant from time to time and place to place, it does change its clothes, age, sex, class, race, and appearance. For example, opiate addiction in nineteenth century North America was predominantly Asian- and Euro-American, but became more black- and Hispanic-American during the twentieth century. An evening pipeful of tobacco among mature men, a tradition among Thai-Lao peasants, has given way to all-day-long cigarette smoking among youth and town dwellers. Skid rows have declined in cities with large Irish populations, only to reappear in Sioux City, Gallup, and elsewhere on the outskirts of Indian reservations. "Pot parties" among college students in the 1960s have been replaced by beer-drinking "keggers" among junior high students in the 1970s. Women's liberation has not been associated with a lessening of men's substance abuse rates towards those lesser rates of women, but rather a marked increase of women's rates towards those of men.

Drugs and alcohol can serve as symbols of a particular lifestyle. Heroin often represents a risk-taking, "macho," antisocietal lifestyle, whether in Harlem or Hong Kong. Social factors offer more lucid explanations for this lifestyle than does the pharmacology of heroin, a calming sedative drug. Youths' illicit substance use probably communicates less about intoxicant effect per se than about attitudes towards family, society, authority, and/or themselves.

SOCIAL PREVENTION AND INTERVENTION OF SUBSTANCE ABUSE

Social Mores and Prevention

All of us influence the behavior of others, and we are influenced by others. This influence is exerted within the family and the extended kin group: among friends, neighbors, and co-workers, and even in crowds. It can be a potent means for preventing substance abuse.

Societies with long exposure to alcoholic beverages have developed means for reducing or even virtually eliminating alcohol-related problems at the family level—and often without recourse to written laws or law enforcement. For example, Chinese and Jewish people have accomplished this by teaching people to learn social drinking behavior at a relatively young age, in a family context, at a ritual or ceremonial time. Alcohol use has become an integral part of the culture, and there are strong social taboos against its use in deviant ways. At the other end of the spectrum, many Arab groups have totally proscribed any use of alcohol. Neither the "childhood training" nor the "prohibition" method is an absolute guarantee of avoiding substance abuse, however. As extended family and religion have been underminded in secular or multiethnic societies, alcoholism has appeared among such groups.[80–82]

The "childhood training" method can be effective for only one or a few psychoactive substances. As a case in point, Jewish people rarely have alcohol-related problems,[83] but do appear in clinical reports with psychiatric problems related to drug use[84] and in surveys indicating liberal use of psychoactive compounds.[85] Moslem peoples, although having little alcohol abuse, sometimes show high rates of heavy cannabis use and/or opiate addiction (as in Afghanistan and Malaysia). Thus, training in social use of alcohol or abstention from alcohol does not "immunize" an individual from using and subsequently abusing other drugs. The "prohibition" approach for all psychoactive substances appears to work for those who remain within the enculturation group; those who leave it are at risk to alcoholism.

Social Mores and Early Intervention

In peasant, tribal, and other stable societies there are traditions by which the group (whether extended kin, work group, or neighbors) exerts its influence on individuals with problematic behavior. A respected leader, group of elders, or family conclave speaks to the person, pointing out past errors and suggesting a new way. If the disturbing behavior continues, sanctions are implemented, such as ignoring the person or withdrawing material support. If the behavior proves a risk to life and property, the person may be exiled or even killed.[86,87]

In highly mobile, relatively anonymous urban societies, these means for influencing others tend to weaken or atrophy. Why be concerned about an intoxicated neighbor when neither party is apt to be living next door a year or two hence? Is it not a waste of time to become involved with a co-worker's alcoholism if one is not getting paid to worry about that? Or what is the use of addressing a relative's substance abuse when family reunions take place only once or twice a year. Unless these questions can receive a positive response, intervention into substance abuse is apt to occur either too late or not at all. For early intervention to be successful, people must be concerned about other people. Perhaps the recent energy crisis and falling birth rate will facilitate this process by engendering greater geographic stability, with subsequent commitment to relatives, neighbors, and co-workers. If so, these economic and demographic changes could be more effective than laws, corporate prevention programs, or treatment facilities in reducing substance abuse.

For early intervention to become widespread, two concepts must be redefined in our society: 1) each person's relation to society, and 2) confrontation. First, our cathexis has been strongly oriented to individualism, a narcissistic independence, the "I" experiences—at the expense of conformity, social responsibility, or "we" experiences. In some respects, this emphasis has resulted in great economic advances, but often at great social cost. This focus must shift if people are to become more concerned with "my community" than "myself." And second, confrontation has been viewed as an angry, rejecting transaction between people—probably due to our mania for responsibility to and for the self alone. Confrontation need involve neither anger nor rejection. On the contrary, it can involve considerable caring, pathos, love, and fear. It is an effort to keep the person inside the group, to keep a relationship intact, to heal and reconcile.

Religion

Individuals rarely engage simultaneously in both active religious practice and substance abuse, whether the religion be animism, Buddhism, Islamism, Judaism, or Christianity. Since organized religions (like most institutions) tend to change slowly and often take reactionary stances, they have not proven a strong resource ot many people facing the rapid changes within the industrial and atomic societies over the last century. Into this vacuum have come a variety of religious movements: neo-Buddhist cults in Asia,[88] youth cults in the United States,[40] charismatic movements among Christians, syncretic religions among Japanese and American Indians.[89,90]

Most recent movements and cults have taken strong stands on psychoactive compounds. Some have ministered primarily to substance abusers, such as the Salvation Army, "Jesus freak" groups, and a charismatic movement in Thailand.[44] Others have forbidden use of alcohol and virtually all self-administered drugs, such as the Mormons and certain youth cults. A few prescribe use of a specific psychoactive substance as a sacramental, such as peyote in the Native American Church, while discouraging other substances, such as alcohol.

Some Protestant groups have facilitated self-help groups, such as Alcoholics Anonymous, Alanon, and Alateen, by providing meeting places in church basements and parish houses. A few Catholic churches and Judaic temples have recently begun to do the same. Perhaps these efforts forecast a greater involvement and more active leadership by organized religion in substance-abuse problems.

Law

Since many legislators, governors, presidents, and bureaucratic directors are trained as attorneys—or are themselves more-or-less law-abiding even if not attorneys—they tend to see law or legislative action as a means for remedying social problems. Since law has proven highly effective in solving numerous fiscal, social, and even health problems, the motivation to use law in this way is particularly strong. But the application of law for regulating substance use and abuse encounters numerous obstacles.

One of these is the nature of chemical dependence itself. Many people dependent on alcohol, opiates, and other drugs are not dissuaded by laws. By the thousands and hundreds of thousands, they break laws to continue their quest for the intoxicant experience.

Prohibition laws, when not accompanied by reeducation and other massive social change, have often caused more problems than they ameliorated. Prohibition in the United States during the 1920s and 1930s did reduce the death rate from alcoholic cirrhosis, but at a social cost in increased criminality and corruption of politicians, government workers, and police that the society could not bear. Pittman[91] has made the point that even our narcotic laws have acted as a "prohibition" strategy against the people who had been legally using narcotic drugs earlier in this century. Prohibition laws against opium in several countries of Asia have only served to replace traditional opium smoking with the problems of heroin addiction.[33] Illicit Irish distillation served as a symbol of dissent to English law for centuries[92]—analagous to Sicilian-Americans in prohibition-era Chicago, heroin-dealing Black Americans in New York's Harlem, and Shan poppy farmers in Burma. Trout[93] has recommended that "continued education and research" and "timely regulatory changes" made "dispassionately outside the political arena," replace impulsive legislation as a means for resolving or reducing our drug-abuse problems.

Laws based on poor pharmacology have done little to maintain respect for law. Inclusion of cannabis as a "narcotic drug" in American law—stimulated in large part by fear of cannabis use among Chicanos[94] —probably injured the cause of law as a means for controlling drug use in the United States.

Despite limitations, law can be effective as a means for restricting or greatly reducing substance abuse. The Aztec nation in pre-Columbian Mexico has clearly defined and strictly enforced rules for alcohol use which seem to have been effective.[95] Several governments of Asia have reduced or eliminated narcotic addiction,[31] and a few Moslem nations have accomplished this with alcohol. But to accomplish regulation there must be a nation-wide integration of virtually all social institutions: law enforcement, religion, health, education, ethnic enclaves, and special interest groups. And the social will must exist to uniformly apply harsh restrictive measures or punishments on all traffickers and recidivists. For many nations the social costs of draconian measures are excessive.

Economics

In particular instances the prevalence of substance abuse cannot be divorced from economic realities. Poppy growers in Asia generally have very high rates of opiate addiction, but no other cash crop can presently substitute for the opium poppy.[12] Irish and Scottish distillers have traditionally had to rely on converting their agricultural products to alcohol in order to survive economically.[92] Recent influx of cannabis and cocaine into the United States has depended on the willingness of bright, daring, often well-educated young people who have had limited employment opportunities within the law. Unless the economic necessities favoring such occupations can be addressed, it is not likely that these activities will abate.

Pathoplastic Nature of Substance Abuse

Even if substance abuse were to be eliminated or largely reduced, the millenium would still not have arrived. Affective illness and family discord occur frequently among people who have recovered from chemical dependency, as well as among their relatives. It is likely that these problems will still occur and perhaps increase if substance abuse is reduced, and they will require professional help.

If primary prevention programs were implemented, such as those employed in several countries of Asia, there would also be social costs. Individual freedoms would be hedged; traffickers and recidivist addicts would be incarcerated for prolonged periods; law enforcers would probably become brutalizers as they often are in coercive governments—whether of the right or the left. Religious and informal social interventions might preclude either this coercive approach or the continuing spread of substance-abuse problems in the United States and much of the world.

REFERENCES

1. Fish MS, Horning EL: Studies on hallucinogenic snuffs. *J Nerv Ment Dis* 124:33–37, 1956.
2. Furst PT, Coe MD: Ritual enemas. *Natural History* 86:88–91, 1977.

3. Yarbrough CC: Therapeutics of cola. *JAMA* 32:1148–1149, 1899.
4. Getahun A, Krikorian AD: Coffee's rival from Harar, Ethopia. *Economic Botany* 27:353–389, 1973.
5. Jarvik ME: Drugs and sexual functioning. *J Fam Pract* 4:944–1006, 1977.
6. Goddard D, deGoddard SN, Whitehead PC: Social factors associated with coca use in the Andean region. *Int J Addict* 4:577–590, 1969.
7. Hanna JM: Coca leaf use in Southern Peru: Some biosocial aspects. *American Anthropologist* 76:281–286, 1976.
8. Burton-Bradley BG: Papua and New Guinea transcultural psychiatry: Some implications of betel chewing. *Med J Aust* 2:744–746, 1966.
9. Carstairs GM: Daru and bhang: Cultural factors in the choice of intoxicant. *Q J Stud Alcohol* 15:220–236, 1954.
10. Swed JF: Gossip, drinking, and social control: Consensus and communication in a Newfoundland parish. *Ethnology* 5:434–441, 1966.
1. Csikszentmihalyi M: A cross-cultural comparison of some structural characteristics of group drinking. *Hum Dev* 11:201–216, 1968.
12. Westermeyer J: Use of alcohol and opium by the Meo of Laos. *Am J Psychiatry* 127:1019–1023, 1971.
13. Simmons OG: The sociocultural integration of alcohol use: A Peruvian study. *Q J Stud Alcohol* 29:152–171, 1968.
14. Blum RH, Blum EM: Drinking practices and controls in rural Greece. *Br J Addict* 60:93–108, 1969.
15. Carney MWP, Lawes TGG: The etiology of alcoholism in the English upper social classes. *Q J Stud Alcohol* 28:59–69, 1967.
16. Ahlstrom-Laakso S: *Drinking Habits among Alcoholics: Finnish Foundation for Alcohol Studies.* New Brunswick, NJ, Rutgers Center for Alcohol Studies, 1975.
17. Bourke JG: Distillation by early American Indians. *American Anthropologist* 7:297–299, 1894.
18. Westermeyer J: The drunken Indian: Myths and realities. *Psychiatric Annals* 4:29–35, 1974.
19. Miller L: The epidemiology of drug abuse in Israel (with special reference to cannabis). *Isr Ann Psychiatry* 10:225–231, 1972.
20. Buchan T: Some aspects of drug abuse in Rhodesia. *Cent Afr J Med* 21:235–238, 1975.
21. Biener K: Drug abuse among Swiss youth. *JAMA* 233:374, 1975.
22. Sargent MJ: Changes in Japanese drinking patterns. *Q J Stud Alcohol* 28:709–722, 1967.
23. Westermeyer J, Soudaly C, Kaufamn E: An addiction treatment program in Laos: The first year's experience. *Drug Alcohol Depend* 3:93–102, 1978.
24. Ogan E: Drinking behavior and race relations. *American Anthropologist* 68:181–188, 1966.
25. Heath DB: Peasants, revolution, and drinking: Interethnic drinking patterns in two Bolivian communities. *Human Organization* 30:179–186, 1971.
26. Gallagher OR: Drinking problems among the tribal Bihar. *Q J Stud Alcohol* 26:617–628, 1965.
27. Ewing JA: Students, sex and marijuana. *Medical Aspects of Human Sexuality* 6:101–117, 1972.
28. Khalifia AM: Traditional patterns of hashish use in Egypt, in Rubin V (ed): *Cannabis and Culture.* Hague, Mouton Publishers, 1975.
29. Slotkin JS: The peyote way. *Tomorrow* 4:64–70, 1955–1956.
30. Nason JD: Sardines and other fried fish: The consumption of alcoholic beverages on a Micronesian island. *J Stud Alcohol* 36:611–625, 1975.

31. Lowinger P: The solution to narcotic addiction in the People's Republic of China. *Am J Drug Alcohol Abuse* 4:164–178, 1977.

32. Musto D: *The American Disease: Origins of Narcotic control.* New Haven, Yale University Press, 1973.

33. Westermeyer J: The pro-heroin effects of anti-opium laws in Asia. *Arch Gen Psychiatry* 33:1135–1139, 1974.

34. Klausner SZ: Sacred and profane meanings of blood and alcohol, in Rubin V (ed): *Cannabis and Culture.* Hague, Mouton Publishers, 1975.

35. Washburn C: Primitive religion and alcohol. *International Journal of Comparative Sociology* 9:97–105, 1968.

36. Leacock S: Ceremonial drinking in an Afro-Brazilian cult. *American Anthropologist* 66:344–354, 1964.

37. Jay EJ: Religions and convivial uses of alcohol in a Gond village in Middle India. *Q J Stud Alcohol* 27:88–96, 1966.

38. Furst PT: *Flesh of the Gods: The Ritual Use of Hallucinogens.* New York, Praeger Publications, 1972.

39. Wellman JF: North American Indian rock art and hallucinogenic drugs. *JAMA* 239:1524–1527, 1978.

40. Hendricks GL: The Jesus movement in campus dress. *Counseling and Values* 17:18–26, 1972.

41. Westermeyer J, Walzer V: Drug abuse: An alternative to religion? *Dis Nerv Sys* 36:492–495, 1975.

42. Heath DB: Drinking problems of the Bolivian Camba. *Q J Stud Alcohol* 19:491–508, 1958.

43. Kearney M: Drunkenness and religious conversion in a Mexican village. *Q J Stud Alcohol* 31:248–249, 1970.

44. Westermeyer J: Nonmedical detoxification at a Buddhist temple in Thailand: Description and evaluation. *J Drug Issues,* 10:221–228, 1980.

45. Doughty PL: The social uses of alcoholic beverages in a Peruvian community. *Human Organization* 30:187–197, 1971.

46. Holmberg AR: The rhythms of drinking in a Peruvian coastal mestizo community. *Human Organization* 30:198–202, 1971.

47. Hough R: Annals of the sea. *The New Yorker,* Aug 2, 1966, p 71.

48. Chopra GS: Sociological and economic aspects of drug abuse in India. *Int J Addict* 7:67–73, 1972.

49. DeLeon G, Wexler HK: Heroin addiction: Its relation to sexual behavior and sexual experience. *J Abnorm Psychol* 81:36–38, 1973.

50. Everett GM: Effects of amyl nitrite ("poppers") on sexual experience. *Medical Aspects of Human Sexuality* 6:146–151, 1972.

51. Goode E: Sex and marijuana. *Sexual Behavior* 2:45–51, 1972.

52. Westermeyer J: Options regarding alcohol usage among the Chippewa. *Am J Orthopsychiatry* 42:398–403, 1972.

53. Bell DS, Trethowan WH: Amphetamine addiction and disturbed sexuality. *Arch Gen Psychiatry* 4:74–78, 1961.

54. Siassi I, Crocetti G, Spiro HR: Drinking patterns and alcoholism in a blue-collar population. *Q J Stud Alcohol* 34:916–926, 1973.

55. Modlin HC, Montes A: Narcotic addiction in physicians. *Am J Psychiatry* 121:358–365, 1964.

56. Vaillant G, Sobowale NC, McArthur C: Some psychological vulnerabilities of physicians. *N Engl J Med* 287:372–375, 1972.

57. Westermeyer J: Alcohol, medications and other drugs, in Nadelson C, Marcotte DB (eds): *Treatment Interventions and Human Sexuality.* New York, Plenum Press, 1980.

58. Woodside M: The first 100 referrals to a Scottish drug addiction treatment centre. *Brit J Addict* 68:231-241, 1973.
59. Blizard PJ: The social rejection of the alcoholic and the mentally ill in New Zealand. *Soc Sci Med* 4:513-526, 1970.
60. Dumont MP: Tabern culture: The sustenance of homeless men. *Am J Orthopsychiatry* 37:938-945, 1971.
61. Feldman J, Su WH, Kaley MM, et al: Skid row and inner-city alcoholics: A comparison of drinking problems and mental problems. *Q J Stud Alcohol* 35:565-576, 1974.
62. Sutter AS: Worlds of drug use on the street scene, in Cressy DR, Ward DA (eds): *Delinquency, Crime and Social Process.* New York, Harper & Row, 1969.
63. Finestone H: Cats, kicks and color. *Social Problems* 5:3-13, 1957.
64. Finestone H: Narcotics and criminality. *Law and Contemporary Problems* 22:69-85, 1957.
65. Agar M: Selecting a dealer. *American Ethnologist* 2:47-60, 1975.
66. Cavan S: *Liquor License: An Ethnography of Bar Behavior.* Chicago, Aldine, 1966.
67. Westermeyer J: Opium dens: A social resource for addicts in Laos. *Arch Gen Psychiatry* 31:237-240, 1974.
68. Thompson T: Drugs as reinforcers: Experimental addiction. *Int J Addict* 3:199-206, 1968.
69. Westermeyer J, Bush J, Wintrob R: A review of the relationship between dysphoria, pleasure, and human bonding. *J Clin Psychiatry* 39:415-421, 1978.
70. Westermeyer J: *Primer on Chemical Dependency: A Clinical Guide to Alcohol and Drug Problems.* Baltimore, Williams & Wilkins, 1976.
71. Cushman P: Relationship between narcotic addiction and crime. *Federal Probation* 38:38-43, 1974.
72. Lindeluis R, Salum I: Alcoholism and crime: A comparative study of three groups of alcoholics. *J Stud Alcohol* 36:1452-1457, 1975.
73 . Piotrowski KW, Losacocco D, Guze SB: Psychiatric disorders and crime: A study of pretrial psychiatric examinations. *Dis Nerv Sys* 37:309-311, 1976.
74. Westermeyer J, Peng G: Opium and heroin addicts in Laos: I. A comparative study. *J Nerv Ment Dis* 164:346-350, 1977.
75. Westermeyer J, Peng G: Opium and heroin addicts in Laos: II. A study of matched pairs. *J Nerv Ment Dis* 164:351-354, 1977.
76. Berger LJ, Westermeyer J: World traveler addicts in Asia: II. Comparison with "stay at home" addicts. *Am J Drug Alcohol Abuse* 4:495-503, 1977.
77. Greden J: The tea controversy in colonial America. *JAMA* 236:63-65, 1976.
78. Halikas JA, Darvish HS, Rimmer JD: The black addict: I. Methodology, chronology and addiction, and overview of the population. *Am J Drug Alcohol Abuse* 3:529-543, 1976.
79. Koon LH, Chuan PS, Ganderia B: Ventilatory capacity in a group of opium smokers. *Singapore Med J* 11:75-79, 1970.
80. Moore R: Alcoholism in Japan. *Q J Stud Alcohol* 25:142-150, 1964.
81. Singer K: Drinking patterns and alcoholism in the Chinese. *Br J Addict* 67:3-14, 1972.
82. Globetti G: Teenage drinking in a community characterized by prohibition norms. *Br J Addict* 69:275-279, 1973.

83. Keller M: The great Jewish drink mystery. *Br J Addict* 64:287–396, 1970.

84. Rosenbloom JR: Notes on Jewish addicts. *Psychol Rep* 5:769–772, 1959.

85. Parry HJ: Use of psychotropic drugs by U.S. adults. *Public Health Rep* 83:799–810, 1968.

86. Westermeyer J: Assassination in Laos: Its psychosocial perspectives. *Arch Gen Psychiatry* 28:740–743, 1973.

87. Westermeyer J: Assassination and conflict resolution in Laos. *American Anthropologist* 74:123–131, 1973.

88. Keyes CF: Millenialism, Theravada Buddhism, and Thai society. *J Asian Studies* 36:283–302, 1977.

89. Kehoe AB: The ghost dance religion in Saskatchewan, Canada. *Plains Anthropologist* 13:269–304, 1968.

90. Lewis TH: The heyoka cult in historical and contemporary Ogalala Sioux society. *Anthropos* 69:2–32, 1974.

91. Pittman D: Transcultural aspects of drinking and drug usage. *Twenty-ninth International Congress on Alcoholism and Drug Dependence.* London, Butterworths, 1971, pp 56–68.

92. Connell KH: Illicit distillation: An Irish peasant industry. *Historical Studies of Ireland* 3:58–91, 1961.

93. Trout ME: Drug abuse: National and international control. *J Legal Med* March/April, 44–48, 1973.

94. Musto DF: The marijuana tax act of 1937. *Arch Gen Psychiatry* 26:101–108, 1972.

95. Paredes A: Social control of drinking among the Aztec Indians of Meso-America. *J Stud Alcohol* 36:1139–1153, 1975.

3 The Pharmacology of Addictive Drugs: A Review of Similarities and Differences

Kenneth Blum
Arthur H. Briggs
Karl Verebey

INTRODUCTION

Numerous attempts have been made to define the loci of the behavioral and tolerance/dependence-producing properties of opiates and ethanol. However, caution must be used in interpreting these results because of the complex interrelationships among various neural elements and their respective neurotransmitters. Thus, while a particular agent is considered to affect only one system, and while biochemically that may indeed be true, the loss or increased stimulation of that specific system may either inhibit or excite several others. Nevertheless, studies of this nature are extremely important in elucidating not only the neurochemical and behavioral mechanisms of opiates and ethanol, but also in unraveling the complex interrelationships that exist between them in the nervous system.

The widespread abuse of both opiates and ethanol in modern societies has resulted in widespread scientific investigation into the mechanisms of action of these two diverse classes of drugs. Although it

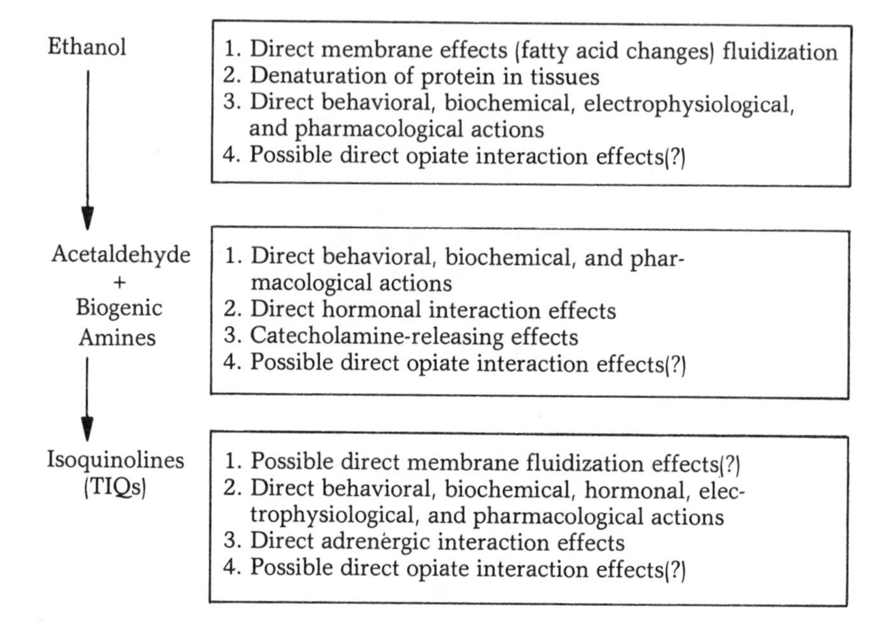

Figure 3-1 Schematic of delineated actions of ethanol, acetaldehyde, and isoquinolines.

is well known that one drug is often associated with concurrent abuse of the other,[1] little effort has been devoted to establishing a common underlying biochemical mechanism. This is not surprising, as it is difficult to envision the biochemical, physiological, or metabolic pathways that complex phenanthrene-type alkaloids would have in common with a simple 2-carbon ethanol molecule. Nevertheless, scrutiny of the literature points towards possible common mechanisms for certain pharmacological actions of these two highly addictive substances.

As a note of caution, generalizations in the interpretation of results derived from interactions between alcohol and narcotic antagonists could be misleading. For example, a significant alteration of an alcohol-induced response (pharmacological, biochemical, physiological, etc) by a narcotic antagonist does not universally indicate that alcohol is acting solely through opiate-mediated sites. Nor would the converse, that is, no similar mechanisms, be true if no significant alteration is found with a narcotic antagonist for a specific effect of alcohol in a biological system. Narcotic antagonism may work only on certain pharmacological actions of alcohol, but not on others. In fact, we suggest that in designing common mechanism studies, use of a narcotic and/or other drugs must be an integral part of the experiment so that direct comparisons can be made.

Another important concept to consider is that ethanol, unlike morphine, may have many varied actions in both a specific and/or nonspecific manner. Morphine for the most part acts by binding to certain specific tissue sites, whereas ethanol may act in part through similar binding sites or contiguous sites, but could act nonspecifically as well.

To illustrate this point Figure 3-1 reveals that the pharmacological, biochemical, and electrophysiological properties of ethanol may be quite specific to either ethanol by itself; acetaldehyde, which is its normal metabolite; or isoquinolines (TIQs), a by-product formed following ingestion of ethanol.[2,3]

NEUROCHEMICAL SIMILARITIES

Due to space limitations a complete review is not presented. However, the advanced or interested reader should refer to books by Loh and Ross,[4] Herz et al,[5] Costa and Trabucchi,[6] and Blum.[7] The available information leads to the following conclusions:

1. Both opiates and ethanol produce alterations in neuroamine metabolism and concentrations. These effects occur after acute administration, during chronic administration, and also during the withdrawal phase.

2. From the results reported, there is little agreement concerning the effect that ethanol or opiate intoxication and withdrawal produces on neurochemical mechanisms. The conflicting results obtained are most likely related to a combination of factors in the experimental protocols, including different routes of administration, different monoamine measuring techniques and/or the timing of the measurements, or the differences in the utilized doses to produce the desired effect (acute intoxication or dependence). While by no means exclusive, this list does point out several reasons for the lack of general agreement concerning the neurochemical alterations elicited by either ethanol or opiates. The lack of agreement obviously precludes a definite conclusion concerning similarities or differences in the neurochemical alterations induced by acute or chronic treatment with either of these substances.

3. Further evidence for a common mechanism of action has been obtained through pharmacological manipulation of central monoaminergic function. Considerable research has implicated changes in brain biogenic amines induced by opiates and ethanol administration as mediating at least some of the effects of acute or chronic drug intoxication. Although the exact mechanism and direction of changes in biogenic

amines induced by ethanol and opiates remain controversial, the fact that changes do occur is generally accepted. This has prompted several investigators to examine the effects of biogenic amines on behavioral changes induced by these two drugs. As with the effects of administration of opiates or ethanol on endogenous neurochemical mechanisms, the effects of prior manipulation of these neurochemicals on subsequent tolerance to, dependence on, and withdrawal from these agents is similarly confusing. The evidence reveals that considerable controversy remains surrounding the neurochemical effects and determinants of both opiates and ethanol. A definitive case cannot yet be made for any single biogenic amine, ion, or other endogenous substance as the initiator of dependence-producing properties of either agent.

4. Some distinct differences are evident in some reports. For instance, while dopamine has been reported to ameliorate ethanol-induced withdrawal,[8] exacerbation of morphine withdrawal has been suggested to occur with stimulation of dopamine receptors.[9] However, in rats, this exacerbation of withdrawal pertained primarily to "dominant" withdrawal signs (eg, jumping), whereas "recessive" signs (eg, diarrhea) decreased.[10] Thus, the definition of exacerbation or amelioration depends on the importance ascribed to particular withdrawal signs, and a conclusion concerning the effects of a drug on abstinence from either opiates or ethanol rests on this definition.

Significant contributions have been made recently, and special emphasis will be devoted to the concept of common mechanisms, such as the interaction of ethanol, TIQs, and opiate-ligand effects.

CLINICAL PHARMACOLOGY

The question of a unified theory of addiction as proposed by Davis and Walsh,[11] Blum,[7] Hamilton et al,[12] and Marshall and Hirst[13] implies that a common biological mechanism engenders an individual to be particularly susceptible to an opiate alkaloid and alcohol.[14] History will finally resolve whether genetic processes are influential on or responsible for drug-seeking behavior for drugs such as ethanol and heroin. Nevertheless, a moderate amount of evidence is beginning to accumulate to support this view,[7,15] even though the clinical symptoms of alcohol and narcotics are not necessarily identical with respect to the addictive process, eg, withdrawal, tolerance, and cross-dependence.

BACKGROUND AND OBSERVATIONS

For approximately a decade our laboratory has actively explored the possibility of common sites of action for alcohol and opiate-like substances. The basis for this investigative search was the hypothesis provided by Davis and Walsh[11] in their controversial paper that appeared in *Science*. They suggested that isoquinolines, eg, tetrahydropapaveroline, the alkaloid-like condensation products formed as a consequence of ethanol metabolism, might be involved in the process of becoming dependent on alcohol. Their speculation was not without an experimental basis, and in fact they succinctly pointed out that benzylisoquinoline alkaloids are requisite intermediates in the biosynthesis of morphine in the poppy plant *Papaver somniferum*.[16] The biogenesis of these alkaloids in mammalian tissues stimulated the speculation that common biochemical mechanisms may exist between opiates and ethanol. In the same year, Cohen and Collins[17] proposed that single TIQs might also be responsible for the acute and chronic effects of ethanol intoxication. This second notion stimulated work on the simpler TIQ, salsolinol. After approximately seven years of collecting data, we suggested the "link" hypothesis, which states that the isoquinolines, when formed following ethanol ingestion, function as an opiate and thus serve as the biochemical "link" between alcohol and opiates.[18]

Explorations of alcoholism and opiate addiction in humans also point toward a possible relationship between alcohol and narcotics. Examples of similarities in human behavior include the finding by Jackson and Richman[19] that many narcotic addicts would use alcohol when drugs such as heroin were not available. In New York City during the period from 1950 to 1962 one-tenth of the deaths among narcotic addicts were attributed to the combined use of narcotics and alcohol,[20] and Baden[21] reported that more than one-fifth of the heroin addicts who died in New York City showed evidence of alcohol abuse. Addicts participating in methadone maintenance programs also reported concomitant heavy use of alcohol.[22,23] Heroin addicts maintained on prolonged methadone treatment and secondarily dependent on alcohol develop a ten-fold increase in mortality over those who were on methadone maintenance alone.[24] In this regard, Casella et al[25] found that ethanol chronically administered to monkeys receiving L-alpha acetylmethadol (LAAM) did not impair the metabolism of LAAM. This work is not in agreement with the finding that acute administration of ethanol resulted in increased brain and liver concentrations of methadone at two time periods and decreased biliary output of pharmacologically active methadone measured by radioactive tracer technique.

Dr Vincent Dole (personal communication to K. Blum) pointed out that in unpublished studies with Dr Kreek, the total amount of ethanol consumed was not significantly altered in high-dose methadone patients when the euphoric properties of heroin were blocked. He argues that if alcohol-induced euphoria is mediated via opiate-like mechanisms, one would expect a marked reduction in alcohol consumption in this specific population. It is agreed that this thesis is very plausible; however, it is possible that chronic alcoholics do not use alcohol solely for euphoric reasons. In fact, there is clinical evidence to suggest that the alcoholic may use alcohol for dysphoric rather than euphoric reasons. If issues like this are kept in mind, we may find the actual relative importance of common to uncommon sites occupied by both alcohol and opiates.

THE "EUPHORIA" QUESTION

The effects of alcohol have often been compared to opiates especially, for the ability of both of these substances to cause analgesia, euphoria, tranquilization, as well as similar neurochemical and physiological effects.[1,26,27] Wikler et al[28] reported that a 60-ml dose of 95% ethanol in man raises the pain threshold approximately 35% to 40%, while not altering other sensory perceptions. Euphoric activity might be the principal property of ethanol that causes dependence liability but not necessarily the only one. The euphoria is a "high," manifested by a diminution of anxiety resulting in a positively reinforcing state of relaxation and physical activity. The paradoxical question has long been debated: is alcohol a stimulant or a depressant? Alcohol is not a stimulant according to a respected textbook of pharmacology and, like other general anesthetics, it is a primary and continuous depressant of the central nervous system. The mechanism of action for the stimulating effect of alcohol is explained through a depressant mechanism. The theory is that the activity of certain inhibitory centers in the CNS is depressed by low doses of alcohol and that this disinhibition results in a "false sense" of stimulation.[26]

This simple explanation, although logical and possibly partially true, has certain shortcomings. First, it presumes that inhibitory neurons are more sensitive to alcohol than are other neurons. Second, it is not clear how the inhibition of the inhibitory centers causes analgesia and anti-anxiety effects in addition to stimulation. It would be easier to understand the euphoric effects of alcohol if alcohol could be shown to release endogenous, opiate-like substances, thus providing the known opiate pharmacological effects, that is, analgesia and anxiety relief.[27] Perhaps the stimulatory effects of alcohol could be viewed as a counteraction to anxiety-related immobilization. Direct evidence to support this theory presently is not at hand. However, interesting and provocative reports are in the literature to support both direct and indirect interactions between alcohol and endorphins or endogenous opiate-like substances.

Endorphins are oligopeptides produced in the brain of mammals. These peptide neurohormones have been shown experimentally to possess all the pharmacological effects of opiate agonists, such as morphine, heroin, and methadone.[29-32] In this review some experimental data will be examined which implicate the possible relationship between alcohol-related, drug-seeking behavior and endorphins.

Since endorphins interact at opiate receptors and have been shown to 1) produce profound analgesia,[29,30] 2) produce dependence when administered to rodents,[31] 3) be euphoric,[33] and 4) alter opiate abstinence in animals and humans,[34] they may be strong candidates for being biological markers to explain the genetic propensity for drug addiction. In addition, the fact that opiates are primary reinforcers in operant self-administration paradigms suggests that endorphins may play a significant role in the central "reward system."

Goldstein[35] and Loh and Law[36] have speculated that a common pathway of all drugs could involve endorphin and opiate receptors. In simpler terms, it is possible that a preexisting genetic deficiency of endorphins in the brain could lead to drug-seeking behavior and subsequent addiction.

ANIMAL FINDINGS

There are reports on acute interactions between alcohol and narcotic-like compounds in experimental animals. Sinclair et al[37] has shown that 60 mg/kg of morphine given to a rat that drinks alcohol markedly suppresses the animal's intake of a 10% solution of the fluid. This ameliorative action of morphine is manifest whether the animal is given free access to alcohol for one day or one month.[38] In the hamster that prefers a 10% alcohol solution over water, morphine reduces the intake of alcohol to near zero volumes, whereas levorphanol, the inactive isomer, reduces alcohol consumption by only about one-half.[39] Of further interest is that naltrexone, the long-acting narcotic antagonist, which binds to the opiate receptor stereochemically, increases voluntary alcohol consumption.

Ho et al[40] similarly found that morphine suppresses preference in alcohol-preferring mice. In the same study, morphine-treated rats undergoing withdrawal and given the free choice between a sucrose solution and one mixed with alcohol, exhibit preference for the latter solution in the 5% to 11% range. Furthermore, a single injection of methadone produces a similar reduction of voluntary alcohol intake.[41] These authors also found that, during the forced drinking of alcohol following pretreatment with morphine, there is an enhanced abstinence sign in the rat compared to controls without morphine. The finding that morphine markedly suppresses alcohol-induced abstinence in the mouse further suggests that a narcotic may exacerbate the degree of physical dependence on alcohol.

However, Gelfand and Amit[42] found that a rat that prefers and drinks a saccharin-adulterated solution of morphine is not affected by an intraperitoneal injection of 4 ml/kg of 20% alcohol in lessening the intake of morphine. In this regard, taste factors,[43] exposure to morphine, and degree of physical dependence may be determining factors.

Of note is the work of Altschuler et al[44] who demonstrated that naltrexone HCl and naloxone HCl alter ethanol self-administration by rhesus monkeys and its discriminative stimulus properties. Further work by the same group assessed the effects of these two narcotic antagonists in fixed-ratio responding for food reinforcement to determine if the effects applied to all appetitive operant tasks. Neither antagonist caused significant alterations in responding. The authors concluded that narcotic antagonism altered ethanol self-administration and drug discrimination was not due to a generalized depressant effect on all appetitive responding, but suggested a specific interaction between the opiate antagonists and ethanol. The authors, on the basis of these experiments, proposed the possible involvement of the endorphinergic system and are actively researching this possibility.

Other experiments by Lorens and Sainati[45] present evidence for the ability of naloxone, a pure opiate antagonist, to block the excitatory effects of ethanol in the rat: naloxone injected one hour after ethanol administration immediately reversed the excitatory effects of ethanol. The authors suggest the possibility that ethanol releases an endogenous opioid which acts at opiate receptors and enhances locomotor activity as similarly seen after the administration of an opiate agonist. This effect can also be produced by direct electrical stimulation of the lateral hypothalamic centers. The implication of the study is that endorphin release might underlie the positively reinforcing properties of ethanol. If this hypothesis is tenable, opiate antagonists (naloxone or naltrexone) should be able to block alcohol-induced analgesia, euphoria, catatonia, and possibly improve alcoholic stupor by displacing endorphins from opiate receptor sites. A preliminary report on four human alcoholic subjects provided encouraging results.[46] Although the reason that these investigators tried naloxone treatment in alcoholic intoxication did not involve the endorphin hypothesis, the results support it.

The first experiments with narcotic antagonists in ethanol-dependent animals showed that administration of naloxone to ethanol-dependent mice did not precipitate jumping behavior, as it does in opiate-dependent mice.[47] However, other experiments[8] showed that concurrent administration of naloxone during ethanol-vapor exposure significantly attenuates the resultant withdrawal convulsions. Naloxone or naltrexone at 5 mg/kg also can inhibit ethanol narcosis in mice, while higher doses of narcotic antagonists (10 mg/kg) potentiate the narcosis.[8]

Additionally, Blum et al[8] as stated previously have shown that the ethanol-withdrawal syndrome in mice is significantly suppressed by a single injection of 10 mg/kg morphine at the fifth hour after ethanol exposure. Many papers report on the positive interaction of narcotic antagonism and psychopharmacological actions of ethanol, including the work of Middaugh et al[48] on locomotor activity; Bass et al[49] on hyperalgesia; Pack and Ferko[50] on abstinence; Blum et al[1] on dependence production; and Ross et al[51] on calcium depletion in regional brain areas.

AMINE METABOLITE THEORY: A POSSIBLE "LINK" BETWEEN ALCOHOL AND OPIATES

The question of a unified theory of addiction has been raised by several investigators,[7] especially with regard to common alcohol and opiate metabolic links. In vivo and in vitro, acetaldehyde, the highly reactive metabolite of ethanol, can condense with a biogenic amine (more specifically, catecholamines) to form a class of compounds termed tetrahydroisoquinolines (TIQ).[2,11,17] An indoleamine such as 5-hydroxytryptamine (5-HT) may form a tetrahydro-B-carboline (TBC) by a similar condensation reaction with an aldehyde in vitro and in vivo.

The fact that these benzylisoquinoline alkaloids are requisite intermediates in the biosynthesis of morphine in the opium poppy *Papaver somniferum*[7,53,54] provided the basis for the intriguing hypothesis that these materials act as endogenously active factors linking the opiates to ethanol.

TIQs can actually be formed in vivo in the laboratory animal or human after alcohol is administered,[2,3,55,56] or in the human in the absence of previous alcohol exposure.[5,7] It is known that a TBC is also capable of being synthesized in the brain.[58-60] These findings further support the fact that these chemicals may be active in the development of alcoholism.[17,52,61,62]

Since the first report in 1910 when Laidlaw[63] demonstrated that one of the TIQs, tetrahydropapaveroline (THP), possesses a potent depressor effect after intravenous infusion, several other condensation products have been found.[64] Enough is now known of their pharmacological actions to reassess the possibility that the TIQs might be the link among these disparate, dependency-producing substances.

An acceptable linkage hypothesis would require that certain pharmacologic criteria be satisfied. Among these would be 1) the necessary identification of isoquinolines in biologic tissues following acute and/or chronic exposure to ethanol; 2) evidence that the isoquinolines contribute to, or share, ethanol- and opiate-like properties; 3) that such pharmacologic actions be enhanced by common potentiators and reduced by appropriate antagonists; 4) that there be common mechanisms to the action of opiates, ethanol, and/or isoquinoline compounds, 5) that the isoquinolines should display some form of interaction with opiate receptors that may also be shared by opiates and ethanol; and 6) that the TIQs show evidence of the production of tolerance and physical dependence that can be correlated to the ethanol- and opiate-related phenomena.

Our intent is not to describe in detail all of the relevant experiments in this area since a more detailed review has been published elsewhere,[18] but rather to organize the available data and point out salient features and significant findings.

Identification

There is evidence for the endogenous formation of the TIQ alkaloid salsolinol in alcoholics and normal humans,[57] in pharmacologically manipulated rats during acute ethanol intoxication,[3] and identification by Sandler and co-workers[56] of two dopamine-derived TIQs (salsolinol and THP) in the urine of Parkinson patients receiving L-dopa therapy. Turner et al[55] also found THP in L-dopa-treated rats exposed to ethanol. While these findings are corroborated by our investigations, our laboratory has tentatively identified significant amounts of an O-methyl derivative of salsolinol in the striata of mice exposed to ethanol vapor. For a complete description of these results the reader is referred to the paper by Hamilton et al.[2]

Although there is evidence that supports the formation of isoquinoline alkaloids during ethanol intoxication, further work is warranted. There must be a particular emphasis on the monitoring of TIQ metabolites and especially the formation of endogenous amounts of brain THP.

Ethanol-like Actions

Marshall and Hirst,[13] reported that salsolinol and 3-carboxysalsolinol enhanced the duration of ethanol-induced narcosis in mice. In other studies salsolinol was found to selectively decrease the activity of mice bred for differential alcohol sensitivities.[65] Other work on ICR Swiss mice revealed a profound "immobilization" response which could be potentiated by calcium chloride at a similar dose known to potentiate ethanol narcosis.[18] In addition, there is evidence that convulsions in mice withdrawing from ethanol are influenced by the condensed amine.[8,18]

Ethanol-like effects on a cellular level have been observed, and include salsolinol-induced depletion of regional brain calcium[51] and inhibition of calcium binding to synaptic membranes.

Opiate-like Actions

The findings that morphine attenuates the ethanol withdrawal syndrome[8] and that naloxone inhibits the development of dependence to ethanol,[1] suggest that ethanol or some metabolite may affect the opiate receptor.

A report by Marshall et al[66] evaluated whether TIQs possess opiate-like activity. These experiments have shown that 3-carboxysalsolinol produces analgesia by itself and increases morphine analgesia. Furthermore, Blum et al[8] reported that salsolinol augmented the analgesic response of 5 mg/kg of morphine given peripherally. Salsolinol behaves in some ways as a partial narcotic agonist in that it causes a dose-related depression of twitch height of the guinea pig ileum that does not progress to a complete suppression of these responses. Preincubation of this tissue with naloxone reduced the responses of both morphine and salsolinol. However, unlike the responses elicited by morphine, naloxone did not reverse the salsolinol-induced inhibition of contractions once the effect was initiated. This work has now been confirmed by Blum et al[67] who also showed addictive effects between D-Ala^2-met enkephalin and salsolinol on this tissue.

Other studies that more clearly indicate narcotic-induced antagonism of the acute pharmacologic actions of the TIQs are: salsolinol-induced regional depletion of brain calcium in rats blocked by naloxone,[5] partial reduction of salsolinol-induced inhibition of calcium binding by naloxone,[65] and blockade of 3-carboxysalsolinol-induced analgesia in mice by naloxone,[66] and a weak direct stereospecific interaction of salsolinol and THP binding to opiate receptors in rat brain.[68]

Clearly, there are interesting correlations between properties of ethanol, opiates, and the TIQs, but the confusing appearance of narcotic agonist, potentiation, and antagonist activities by the TIQs demands attention and resolution.

Tolerance and Physical Dependence Production By Isoquinoline Alkaloids

A requisite criterion for the "link" hypothesis must include cross-tolerance between the two addictive agents, ethanol and morphine, the TIQ derivatives, and the ability of the latter to demonstrate physical dependence. Despite the evidence of a common biochemical bond as reported by Ross et al[51] and the appearance of sensitization between ethanol and opiates in mice, these aspects have not been widely examined using TIQs. However, it is important at this point to consider the idea that tolerance production induced by morphine may be due to different mechanisms than the induction of tolerance to ethanol. In fact, there is a consensus from the literature that alcohol and opiates differ in tolerance mechanisms. It is obviously premature to attempt to consider opiate-agonist physical dependence development as being initiated by these amine condensation products; yet they produce interesting responses, and further experimentation is certainly warranted.

In conclusion, certain isoquinolines possess similarities to opiates: 1) they form in the brain after ethanol administration,[2,3] 2) they weakly bind stereospecifically to rat and guinea pig brain opiate receptors,[35,68] 3) they prolong ethanol narcosis,[13] 4) they produce narcosis by itself,[65] 5) they produce analgesia and potentiate morphine analgesia,[66] 6) they inhibit the electrically-induced contractions of the guinea pig ileum which is prevented by the narcotic antagonist naloxone, and 7) they deplete brain calcium, like ethanol and morphine, a response that can be blocked by naloxone.[51] Thus, isoquinolines, because of their ability to act on similar sites or contiguous sites in the brain where opiates act (endorphin receptor sites), may function as a "link" between ethanol and opiates.[18]

INTERRELATEDNESS VIA ENDORPHIN-OPIATE RECEPTOR SYSTEM

Pinsky et al[69] suggested that the central endorphin-opiate receptor system is a likely candidate for a good part of the opiate-alcohol interactions that have been cited here. It may be the vulnerable central site in the process of alcohol consumption in humans.

This section will focus on the interaction of alcohol and oligopeptide opiate (enkephalin) interactions with particular emphasis on behavioral and opiate receptor effects.

Realistic scrutiny must somehow determine the possible similarities or differences that exist between junkies in the street, lying, cheating, stealing, and prostituting themselves to obtain a fix of heroin, and cirrhotic alcoholics who spend a month's pay for alcohol while impinging on their family's lifestyles. In most countries there is impunity from the law in only the latter case, but a similarity may exist when we consider the underlying mechanisms involved in the desire to obtain the addictive agents. It may be possible to set up an animal or human model to test this hypothesis.

Scientists have long recognized that they know of no single model that can be used to study alcoholism with all of its complex characteristics. However, certain aspects could be evaluated individually, which might then be integrated into one entity that simulates the elements that make up alcoholism. This concept is depicted in Figure 3-2 and includes: acute actions, chronic actions, opiate-mediated responses, preference, genetics, and developmental pharmacology. However, since much has already been discussed in this review, additional aspects may be found elsewhere.[67,69]

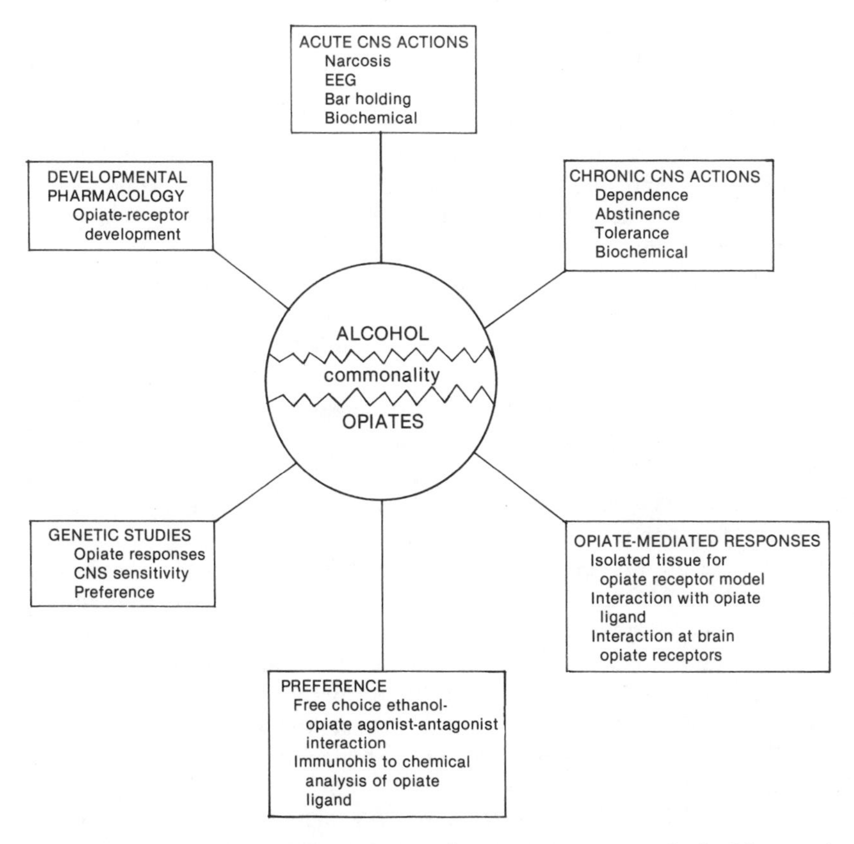

Figure 3-2 Schematic representation of components researched with regard to common mechanism theory of alcohol and opiates.

Our goal is not to review in detail but rather to summarize the general tenor of the relevant findings and interpretations of others, and discuss new findings not reviewed so far. Since genetics may have putative relevance to alcoholism, it has been the subject of several reviews.[70,71]

Scientific Interpretations and Speculations

Considerable evidence suggests possible common territories for both alcohol and opiates, but we must caution against the generalization that these agents are similar in every way and in every action through common mechanisms. Through sophisticated and careful design of experiments we may define the specific interactions which are similar and/or dissimilar for these highly addictive agents. Nevertheless, certain aspects of this problem can be now discussed in a scientific, nonjudgmental framework. The following points in this regard are now presented:

1. The work on the interaction effects of ethanol, salsolinol, and enkephalin on the vas deferens and guinea pig ileum, even though characterizations of the naloxone blockage, is not complete but suggests that common mechanisms may be involved for both opiates and ethanol.

2. Differential sensitivity to bar-holding impairment by ethanol in particular strains of mice correlates well with dependence development potential. This method may be useful as a genetic determinant to brain sensitivity and dependence development to ethanol and central ethanol-opiate interactions. These results suggest a correlation with the human in that a person who may be genetically prone to alcohol abuse due to adaptation may be less sensitive (tolerant) to both the intoxicating- and dependence-producing effects of ethanol. Similar sensitivity differences have also been found for opiates but the data are still controversial and do not allow for a definitive conclusion and comparison to alcohol.

3. A common denominator theory of addiction has been raised in our laboratory[7,18] and by Davis and Walsh,[11] especially with regard to alcohol and opiate common links. As previously mentioned, when alcohol is ingested, a new class of compounds, termed isoquinolines, has been found to form. These isoquinolines, because of their ability to act on similar associated sites (calcium-ATPase) in the brain (opiate receptors), may function as the link between alcohol and opiates. The isoquinolines have also been found to induce long-term drinking of alcohol in rodents. This may be important when we consider that certain mice that are gentically bred to drink alcohol also tend to drink more morphine compared to other mice that do not drink alcohol or

morphine. This finding, although supportive of a common mechanism for preference, has been argued by the work of Whitney and Horowitz[72] who showed differential preference of morphine and ethanol in certain sublines of mice.

4. Since endorphins interact at opiate receptors and have been shown to produce profound analgesia, produce dependence when administered to rodents, be euphorogenic, and alter opiate abstinence in animals and humans, they are strong candidates for being biological markers to explain the genetic propensity for substance addiction. In addition, since opiates are primary reinforcers in operant self-administration paradigms, the endorphins may play a significant role in the central "reward system."

Goldstein[35] and Loh and Law[36] have speculated that a common pathway of all drugs could involve endorphin and opiate receptors. In simpler terms, it is possible that a preexisting genetic deficiency of endorphins in the brain could lead to drug dependence.

Opiate effects are not uniquely related to any of the known neurotransmitters, although changes in content and turnover of several transmitters are associated with opiate actions and with tolerance and dependence. Accordingly, Goldstein points out that the "opiate receptors seem to subserve a general inhibitory function in various neuronal pathways—and that administration of an exogenous opiate might result in suppression of synthesis of endogenous opioids, by analogy to the effects of administering a hormone." This would suggest that administration of an opiate would produce a negative feedback due to saturation of the opiate (endorphin) receptor, which could cause a shut-down of endorphin production. It is tempting to speculate that the ethanol or possibly a metabolite (isoquinoline) saturate opiate (endorphin) receptor sites and also produce a negative feedback inhibition of enkephalin production. This is just a beginning, and future research may lead to a clearer understanding of tolerance, physical dependence, and withdrawal from alcohol and opiates as well as from other addictive agents. Other models that have clinical importance to test this hypothesis of common mechanisms are being developed and actively researched.

5. If it is true that ethanol's reinforcing effects are exerted through interactions with endorphinergic systems, it is not surprising that numerous investigators found similarities in certain opiate and ethanol effects.[1] These similarities in their reinforcing properties may explain why animals are willing to substitute opiates for alcohol and vice versa,[73] and it may explain why naloxone alters self-administration of ethanol in primates[44] and rodents. Both opiates and alcohol directly or indirectly interact with the opiate receptors.[69,74] If this is correct, then in chronic alcoholism or opiate dependence the frequent occupation of

the opiate receptor sites by exogenous opioid causes contact desensitization of the receptors (pharmacological tolerance), and their reinforcing effects are lost unless the dose is increased. Chronic excessive use of both alcohol and opiates will result in the abolition of stimulation of the reward centers in the limbic system, for which effect both drugs were initially consumed. To preserve the reinforcing effects of both drugs, moderate and temporally well-spaced use of these drugs is necessary. The literature indicates that numerous opiate abusers are "chipping" or taking opiates sporadically; they do not develop the street-addict personality.[75,76] The parallel character in the alcohol scene is the light-to-moderate social drinker. Consensus in the literature would argue that an alternative explanation exists for alcohol which is not similar to opiates with regard to tolerance. Tolerance for alcohol may be a membrane-fluidization and alternative-of-fatty-acids problem not at all linked to morphine-induced neuronal adaptation; thus tolerance to alcohol and narcotic-like compounds probably is due to two distinct mechanisms.

6. A proposed central model is derived from available data and indicates that opiates can interact with opiate receptor[74,12] and/or a calcium site, and these actions can be blocked by the narcotic antagonist naloxone; ethanol can also interact at a calcium site, and this can also be blocked, but not reversed, by naloxone[39,51]; isoquinolines can interact with the opiate receptor[68] and naloxone could partially block, but not reverse, opiate interaction effects[12,67]; isoquinolines could interact with the calcium site and be partially blocked by naloxone, but not reversed[65]; ethanol interacts with a nonspecific site (membrane fluidization), which cannot be altered by naloxone[77,78]; and barbiturates and benzodiazepines interact with receptors associated with GABA,[79,80] which is not blocked naloxone. There may be an association between ethanol responses, but exact mechanisms are being actively investigated.[81,82]

SOCIAL ISSUES AND CONCERNS

Problems of opiate and alcohol abuse are recognized only when the dependence becomes overwhelming and interferes with the individual's daily functioning. The question is often asked: Why are these substances used, misused, and abused? If both alcohol (or a metabolite) and opiates act through the opiate receptors or an associated site, the answer may be at hand. Opiates and endorphins are able to divert the subject's attention from emotional as well as physical pain. This characteristic of opiates is responsible for their excellent analgesic properties, highly valued in the treatment of intractable pain.[27] Common stress, related to the varied intensities of life responsibilities, is often interpreted as emotionally painful events. Some people tolerate them while others have major breakdowns. In the future, these differences in individual stress tolerance might be found to be related to endorphin physiology; at present, however, the involvement of endorphins is speculative based on opiate agonist behavioral effects. Subjects under intolerable emotional stress may use alcohol or opiates to escape, while others use nothing and may develop various types and degrees of psychosis. From these observations it would appear logical that the physiological role of endorphins might be to attain psychological homeostasis in the CNS.[83] Endorphins are released from storage vesicles during major stressful events or are chemically released, possibly by alcohol, to provide a defense mechanism during emotional crises or to stimulate a feeling of euphoria. If this role for endorphins can gain experimental support, mental illness characterized by intolerance of stress may well be classified as another metabolic disorder similar in mechanism to diabetes. For insulin deficiency in diabetes, an exogenous insulin is administered; for endorphin deficiency in the treatment of psychosis, an exogenous opiate agonist or a synthetic endorphin may become useful. Furthermore, it is observed that mental illness and opiate and alcohol abuse often begin during these neuroendocrinologically explosive years when some adolescents develop a temporary deficiency of endorphins. If this can be substantiated following diagnosis, an exogenous opiate treatment may bridge the temporary deficiency and perhaps prevent the concomitant development of psychosis or chronic substance abuse.

The choice of the type of drug used by any individual is influenced by both external and internal forces. The external force consists of the cultural background formulated by social mores. For example, Oriental cultures historically were permissive with respect to opium use, which promotes passive adaptation of the individual to the environment. Western cultures are permissive regarding the use of alcohol, which is known to reduce inhibition and expose aggressive behavior. In Oriental cultures the use of alcohol and in Western cultures the use of opiates are considered flagrant abuses. Nevertheless, in both cultures users of the socially unacceptable drugs exist. This directs attention toward the internal forces. It is suggested that an individual will use a drug that fosters a sense of well-being. Thus, the opiate or alcohol user's psychological constitution is a major factor in the choice of drug. According to Pittel,[84] dependence on either alcohol or opiates is likely to occur in individuals who lack the psychological stability to interact with inner structures. Among all drugs, however, opiates have a unique quality to directly satisfy all primary needs.[85] Presumably this is achieved through the stimulation of the "reward" or "pleasure" centers in the limbic system. If the mode of action of endorphins is similar to opiates, it may be substantiated that alcohol and perhaps other euphoria-producing drug effects are achieved through endorphin release or direct interaction with the endorphin receptor sites in the CNS. Although opiate and alcohol dependence appear different superficially, the primary reinforcing effects in both may be related to their ability to stimulate directly or indirectly the opiate receptor.

IMPLICATIONS

If pleasure states induced by drugs are also attainable naturally via the endorphins, should society condemn the self-medicators of exogenous euphorigens when they turn to artificial means? This question is especially poignant if the substance is used to compensate for a genetically deficient or functionally dormant natural endorphin system. The uncritical, hedonistic, and self-destructive abuse of drugs should always be seriously discouraged; however, should drug use be illegal if it means psychological stabilization and functional rehabilitation? In this respect, current drug laws may be inadequate because no recognition is given to possible biological need and/or drug-related benefits.

In essence, heroin, morphine, and methadone-like substitutes are as natural to our physiology as insulin, ACTH, or the estrogens, among others. Excessive amounts or deficiencies of any of these substances lead to pathological states. Thus the opiate-like drugs may not necessarily present more or less danger to humans than any of the hormones mentioned. A major difference between opiates and hormones is that opiates affect behavior more visibly, particularly toward society, than hormones, and that the issue of heroin abuse has become highly political. Much legislation in this area was strictly emotional—not based on scientific evidence. Further elucidation of the role of opioid peptides in human behavior may unravel some of the mysteries of the human brain. Such scientific advances would lead to rational medical, legal, and social approaches to the elimination of substance and alcohol misuse and provide freer access to the beneficial use of euphorigens in therapeutic medicine and psychiatry.

CONCLUSION

At its best, the "common denominator" theory is an interesting working hypothesis which may serve as a model for future directions.

We must await solid experimental evidence in both animals and humans before we can begin to understand the intricate nature of alcohol and opiate dependence, their similarities, and differences.

REFERENCES

1. Blum K, Hamilton MG, Wallace JE: Alcohol and opiates: A review of common neurochemical and behavioral mechanisms, in Blum K (ed): *Alcohol and Opiates: Neurochemical and Behavioral Mechanisms.* New York, Academic Press, 1977, pp 203–235.

2. Hamilton MG, Blum K, Hirst M: Identification of an isoquinoline alkaloid after chronic exposure to ethanol. *Alcoholism Clin Exp Res* 2:133–137, 1978.

3. Collins MA, Bigdeli MB: Tetrahydroisoquinolins in vivo: I. Rat brain formation of salsolinol, a condensation product of dopamine and acetaldehyde under certain conditions during ethanol intoxication. *Life Sci* 16:585–601, 1975.

4. Loh HH, Ross D (eds): *Advances in Biochemical Psychopharmacology: Neurochemical Mechanisms of Opiates and Endorphins.* New York, Raven Press, 1979, vol 20.

5. Herz A, Hollt V, Klee WA, et al: Molecular neurochemistry of active drugs: Group report, in Fishman J (ed): *The Basis of Addiction.* Berlin, Abakon, 1978, vol 8, pp 513–526.

6. Costa T, Trabucchi M: *The Endorphins: Advances in Biochemical Psychopharmacology.* New York, Raven Press, 1978, vol 18.

7. Blum K (ed): *Alcohol and Opiates: Neurochemical and Behavioral Mechanisms.* New York, Academic Press, 1977.

8. Blum K, Wallace JE, Schwertner HA, et al: Morphine suppresion of ethanol withdrawal in mice. *Experientia* 32:79–82, 1976.

9. Iwamoto ET, Ho IK, Way EL: Sudden elevation of brain dopamine after naloxone-precipitated withdrawal in morphine dependent mice and rats. *J Pharmacol Exp Ther* 187:567–588, 1973.

10. Herz A, Blasig J, Papeschu R: Role of catecholaminergic mechanisms in the expression of the morphine abstinence syndrome in rats. *Psychopharmacologia* 39:121–143, 1974.

11. Davis VE, Walsh MJ: Alcohol, amines and alkaloids: A possible basis for alcohol addiction. *Science* 167:1005–1007, 1970.

12. Hamilton MG, Hirst M, Blum K: Opiate-like activity of salsolinol on the electrically stimulated guinea pig ileum. *Life Sci* 25:2205–2210, 1979.

13. Marshall A, Hirst M: Potentiation of ethanol narcosis by dopamine and L-dopa-based isoquinolines. *Experientia* 32:201–203, 1976.

14. Myers RD: Tetrahydroisoquinolines in the brain: The basis of an animal model of alcoholism. *Alcoholism Clin Exp Res* 2:145–154, 1978.

15. Vesell ES, Braude MC: Interactions of drugs of abuse. *Ann NY Acad Sci* 281:1–489, 1976.

16. Battersby AR: Alkaloid biosynthesis. *Q Rev* 15:259–286, 1961.

17. Cohen G, Collins M: Alkaloids from catecholamines in adrenal tissue; possible role in alcoholism. *Science* 167:1749–1751, 1970.

18. Blum K, Hamilton MG, Hirst M, et al: Putative role of isoquinoline alkaloids in alcoholism: A link to opiates. *Alcoholism Clin Exp Res* 2:113–120, 1978.

19. Jackson GW, Richman A: Alcohol use among narcotic addicts. *Alcohol World* 1:25–28, 1973.

20. Helpern M, Rho Y: Deaths from narcotism in New York City. *NY State J Med* 66:2391, 1966.

21. Baden MM: Homicide, suicide and accidental death among narcotic addicts. *Hum Pathol* 3:92–95, 1972.

22. Gearing FR: *Success and Failures in Methadone Maintenance Treatment of Heroin Addiction in New York City, 1970.* U.S. Public Health Service Publication No. 2172, 1971.

23. Wepper RS, Ager MH: Immediate precursors of heroin addiction. *J Health Soc Behav* 12:10–18, 1971.

24. Roizin L: Interaction of methadone and ethanol. Presentation at Meeting of Eastern Section of American Psychiatric Association, New York, 1969.

25. Casella GA, Deneau GA, Mule SJ: The effect of ethanol on the metabolism of chronically-administered 1-alpha-acetylmethadol (LAAM) in the rhesus monkey. *Res Commun Chem Pathol Pharmacol* 21:351–354, 1978.

26. Richie JJ: The aliphatic alcohols, in Goodman LS, Gilman A (eds): *The Pharmacological Basis of Therapeutics.* New York, Macmillan Co, 1975, pp 137–151.

27. Jaffe JH, Martin WR: Narcotic analgesics and antagonists, in Goodman LS, Gilman A (eds): *The Pharmacological Basis of Therapeutics.* New York, Macmillan, 1975, pp 245–283.

28. Wikler A, Goodell H, Wolff HG: Studies on pain: The effects of analgesic agents on sensations other than pain. *J Pharmacol Exp Ther* 83:294–299, 1945.

29. Loh HH, Tseng LF, Wei E, et al: β-endorphin is a potent analgesic agent. *Proc Natl Acad Sci USA* 73:2892–2895, 1976.

30. Graf L, Szekely JI, Ronai ZA, et al: Comparative study on analgesia effect of met-5-enkephalin and related lipotropin fragments. *Nature* 263:240–242, 1976.

31. Wei E, Loh H: Physical dependence on opiate-like peptides. *Science* 193:1262–1263, 1976.

32. Plotnikoff NP, Kastin AJ, Coy DH, et al: Neuro-pharmacological actions of enkephalin after systemic administration. *Life Sci* 19:1283–1288, 1976.

33. Klein NS, Li CH, Lehmann HE, et al: β-endorphin-induced changes in schizophrenic and depressed patients. *Arch Gen Psychiatry* 31:1111–1113, 1977.

34. Catlin DH, Hui KK, Loh HH, et al: β-endorphin: Subjective and objective effects during acute narcotic abstinence in man, in Costa E, Trabucchi M (eds): *The Endorphins: Advances in Biochemical Psychopharmacology.* New York, Raven Press, 1978, vol 18, pp 341–350.

35. Goldstein A: Future research in opioid peptides (endorphins): A preview, in Blum K (ed): *Alcohol and Opiates: Neurochemical and Behavioral Mechanisms.* New York, Academic Press, 1977, pp 397–403.

36. Loh HH, Law PY: Pharmacology of endogenous opiate-like peptides, in Blum K (ed): *Alcohol and Opiates: Neurochemical and Behavioral Mechanisms.* New York, Academic Press, 1977, pp 321–340.

37. Sinclair JD, Adkins J, Walker S: Morphine-induced suppression of voluntary alcohol drinking in rats. *Nature* 246:425–427, 1973.

38. Sinclair JD: Morphine suppresses alcohol drinking regardless of prior alcohol access duration. *Pharmacol Biochem Behav* 2:409–412, 1974.

39. Ross DH, Hartmann RJ, Geller I: Ethanol preference in the hamster: Effects of morphine sulfate and naltrexone, a long acting morphine antagonist. *Proc West Pharmacol Soc* 19:326–330, 1976.

40. Ho AKS, Chen RC, Morrison JM: Opiate-ethanol interaction studies, in Blum K (ed): *Alcohol and Opiates: Neurochemical and Behavioral Mechanisms.* New York, Academic Press, 1977, pp 189–202.

41. Ho AKS, Chen RC, Morrison JM: Potential interactions between narcotics and narcotic antagonists with ethanol. *Ann NY Acad Sci* 281:279–310, 1976.

42. Gelfand R, Amit Z: Effects of ethanol injections on morphine consumption in morphine preferring rats. *Nature* 259:415–416, 1976.

43. Ho AKS: Pharmacological and biochemical studies with beta-carboline analogs. *Curr Dev Psychopharmacol* 4:151–177, 1978.

44. Altschuler HL, Feinhandler D, Aitken C: The effects of opiate antagonist compounds in fixed-ratio operant responding in rats. *Fed Proc* 38:424, 1979.

45. Lorens SA, Sainati SM: Naloxone blocks the excitatory effects of ethanol and chlordiazepoxide on lateral hypothalamic self-stimulation behavior. *Life Sci* 23:1359–1364, 1978.

46. Schenk GK, Enders P, Engelmeier MP, et al: Application of the morphine antagonist naloxone in psychiatric disorders. *Arzneim Forsch* 28:1274–1277, 1978.

47. Goldstein A, Judson BA: Alcohol dependence and opiate dependence: Lack of a relationship in mice. *Science* 172:290–292, 1971.

48. Middaugh LD, Read E, Boggan WO: Effects of naloxone on ethanol induced alterations of locomotor activity in C57BL/6 mice. *Pharmacol Biochem Behav* 9:157–160, 1978.

49. Bass MB, Friedman HJ, Lester D: Antagonism of naloxone hyperalgesia by ethanol. *Life Sci* 22:1939–1946, 1978.

50. Pack RL, Ferko AP: The effects of ethanol and pentobarbital on the naloxone precipitated escape response in morphine dependent mice. *Arch Int Pharmacodyn Ther* 228:58–67, 1977.

51. Ross DH, Medina MA, Cardenas HL: Morphine and ethanol: selective depletion of regional brain calcium. *Science* 186:63–64, 1974.

52. Dajani RM, Saheb SE: A further insight into the metabolism of certain carbolines. *Ann NY Acad Sci* 215:120–123, 1973.

53. Schapf C, Bayerle H: Zurfage der Biogenesese der Isochinolenalkaloid: Die Synthese des 1-methyl-6-7-dioxy 1,2,3,4, tetrahydro-isochinolins inter physiolischen bednigemyen. *Justus Liebigs Ann Chem* 513:190–202, 1934.

54. Pellitier SW: *Chemistry of the Alkaloids.* New York, Van Nostrand Reinhold Co, 1970.

55. Turner AJ, Baker KM, Algeri S: Tetrahydropapaveroline: Formation in vivo and vitro in rat brain. *Life Sci* 14:2247–2257, 1974.

56. Sandler M, Carter SB, Hunter KR: Tetrahydroisoquinoline alkaloids: In vivo metabolites of L-dopa in man. *Nature* 243:439–444, 1973.

57. Nijm WP, Riggin R, Teas G: Urinary dopamine-related tetrahydroisoquinolines: Studies of alcoholics and non-alcoholics. *Fed Proc* 36:334, 1977.

58. Hsu LL, Mandell AJ: *Res Commun Chem Pathol Pharmacol* 12:355–363, 1975.

59. Wyatt RJ, Erdelyi E, DoAmaral RJ, et al: Tryptoline formation by a preparation from brain with 5-methyltetrahydrofolic acid and tryptamine. *Science* 187:853–855, 1975.

60. Hsu LL: *Life Sci* 19:493–496, 1976.

61. Davis VE, Walsh MJ: Effect of ethanol on neuroamine metabolism, in Israel Y, Mardones J (eds): *Biological Basis of Alcoholism.* New York, Interscience, 1971, pp 73–97.

62. Cohen G: Alkaloid products in the metabolism of alcohol and biogenic amines. *Biochem Pharmacol* 25:1123–1128, 1976.

63. Laidlaw PP: The action of some isoquinoline derivatives. *Biochem J* 5:243–273, 1910.

64. Hirst M: Pharmacology of isoquinoline alkaloids and ethanol interactions, in Blum K (ed): *Alcohol and Opiates: Neurochemical and Behavioral Mechanisms.* New York, Academic Press, 1977, pp 167–187.

65. Ross DH: Inhibition of high affinity calcium binding by salsolinol. *Alcoholism Clin Exp Res* 2:139–143, 1978.

66. Marshall A, Hirst M, Blum K: Morphine analgesia augmentation by and direct analgesia with 3-carboyx-salsolinol. *Experientia* 33:745–755, 1977.

67. Blum K, Briggs AH, Elston SFA, et al: Validation of central and peripheral models to evaluate alcohol sensitivity and opiate-membrane interactions as a function of genotype dependent responses in three different strains of mice, in Eriksson K, Sinclair JD, Kiianmaa K (eds): *Animal Models in Alcohol Research.* New York, Academic Press, 1980, pp 85–92.

68. Greenwald JE, Fertel RH, Wong LK, et al: Salsolinol and tetrahydropapaveroline bind opiate receptors in the rat brain. *Fed Proc* 38:379, 1979.

69. Pinsky C, Labella FS, Leybin LC: Alcohol and opiate narcotic dependencies: Possible interrelatedness via central endorphin: Opiate receptor system, in Smith DE, Anderson SM, Buxton M, et al (eds): *A Multicultural View of Drug Abuse. Proceedings of the National Drug Abuse Conference.* Boston, GK Hall & Co, 1978.

70. Lester D, Freed EX: A rat model of alcoholism. *Ann NY Acad Sci* 197:54–59, 1972.

71. Dietrick RA, Collins AC: Pharmacogenetics of alcoholism, in Blum K (ed): *Alcohol and Opiates: Neurochemical and Behavioral Mechanisms.* New York, Academic Press, 1977, pp 109–139.

72. Whitney G, Horowitz GP: Morphine preference of alcohol-avoiding and alcohol preferring C57/BL mice. *Behav Genet* 8:177–182, 1978.

73. Smith SG, Werner TE, Davis WM: Intravenous drug self-administration in rats: Substitution of ethyl alcohol for morphine. *Psych Rec* 25:17–20, 1975.

74. Pert CB, Snyder SH: Opiate receptor: Demonstration in nervous tissue. *Science* 179:1011, 1973.

75. Powell DH: A pilot study of occasional heroin users. *Arch Gen Psychiatry* 28:586–594, 1973.

76. Gay GR, Winkler JJ, Newmeyer JA: Emerging trends of heroin abuse in the San Francisco Bay Area. *J Psychedelic Drugs* 4:53–55, 1971.

77. Littleton JM, John GR, Grieve SJ: Alternations in phospholipid composition in ethanol tolerance and dependence. *Alcoholism Clin Exp Res* 3:50–56, 1979.

78. Johnson DA, Lee NM, Cook R, et al: Ethanol tolerance in reconstituted and intact neuromembranes: A fluorescence study. *Fed Proc* 423:1027, 1979.

79. Ticku MK, Olsen RW: Interaction of barbiturates with dehydropicrotoxin binding sites related to GABA receptor-ionophore system. *Life Sci* 22:1643–1651, 1978.

80. Costa T, Rodbard D, Pert CB: Is the benzodiazepine receptor coupled to a chloride anion channel? *Nature* 277:315–317, 1979.

81. Tiku MK: The effects of acute and chronic ethanol administration and during withdrawal on aminobutyric acid receptor binding in rat brain. *Br J Pharmacol* 70:403–410, 1980.

82. Volicer L: Brain Research Supplement. GABA levels and receptor binding after acute and chronic ethanol administration (Inter Sym on GABA and other Inhibitory Neurotransmitters), Myrtle Beach, SC. Fayetteville NC, Ankao Publishing Co, (in press).

83. Verebey K, Volavka J, Clouet DH: Endorphins in psychiatry: An overview and hypothesis. *Arch Gen Psychiatry* 35:877–888, 1978.

84. Pittel SN: Psychological aspects of heroin and other drugs. *J Psychedelic Drugs* 4:40–45, 1971.

85. Wikler A: A psychodynamic study of a patient during experimental self-regulated readdiction to morphine. *Psychiat Q* 26:270–293, 1952.

4 Personality and Psychopathology: A Comparison of Alcohol- and Drug-Dependent Persons

Jerome F. X. Carroll

In comparing any two groups, one will nearly always observe *both* similarities and differences. The extent of the observed differences and similarities—and the importance attributed to these observations—unfortunately will often be determined as much by the values, expectations, politics, and biases of the observer as by the similarities and differences that actually exist between any two groups being observed.

The growing body of literature relating to experimenter and therapist bias attests to this generalization.[1-11] In addition, a number of writers have suggested that diagnostic labels, such as alcoholic, addict, and mentally ill, may be very detrimental to the troubled person's recovery.[12-15]

With this caveat in mind, let us begin by stating that *yes* there are differences between "alcoholics" and "addicts" with respect to personality and psychopathology, and these differences do have implications for treatment.[16,17] At the same time, however, many similarities also have treatment implications. The similarities in etiology, dynamics, defenses, psychobiosocial consequences of the substance abuse, and the substance abusers' treatment needs, however, seem to have greater clinical significance than the differences. Others expressing similar observations and conclusions include Driscoll,[18] Driscoll and Barr,[19] Ferneau,[20] Belter,[21] Cohen,[22] and Chafetz.[23]

The differences that do emerge between these two groups of substance abusers appear to be determined far more by demographic variables, such as age, race, sex, and socioeconomic status, and the lifestyle centered around abusing a licit versus an illicit substance than by any pre-addictive personality differences. These differences, moreover, can be used by experienced, skilled therapists to facilitate treatment of mixed groups of addicts and alcoholics.[24] On the other hand, clinicians who are ill-prepared and/or resistive to working with both alcoholics and addicts may very well mismanage such differences.[25,26]

The argument can also be made that most experienced therapists tend to pursue a common course of therapy in assisting individuals with problems-in-living regardless of the theoretical orientation and training of the therapist and the patient's presenting problem.[27] Egan,[28] for example, has outlined a three-stage model of the counseling process which is content-free, that is, it could be applied to alcoholics, addicts, and other troubled people with equal effectiveness.

Before discussing the similarities and differences of personality and psychopathology between alcoholics and addicts as if these two categories of substance abusers existed in some absolute measure, it is also important to note that increasing numbers of substance abusers are now reporting multiple substance abuse.[29-31] These patterns of multiple substance abuse are typically distinguished by the abuse of *both* alcohol and other drugs, thus the distinction between alcoholic and addict appears to be relative rather than absolute and of diminishing importance.

ETIOLOGY, DEFENSES, DYNAMICS, AND THE THERAPEUTIC PROCESS

In considering the etiology of alcoholism and drug addiction, researchers and clinicians generally cite the following factors as root cause(s) of the addiction: disturbed family relationships, parental modeling of substance abuse, social alienation, loneliness, feelings of powerlessness and inadequacy, normlessness, depression and anxiety, poverty, social prejudice, sexual frustration, guilt, repressed feelings of anger and rage, and acts of violence and crime. Obviously, these "causative factors" are commonly found in the backgrounds of *both* alcoholics and drug addicts.

With respect to defenses, *both* alcoholics and addicts employ essentially the same defenses: denial, rationalization, and projection. Both manipulate while in treatment in order to avoid making meaningful changes in their addictive lifestyles. Examples of some of the more commonly encountered manipulative ''games'' of alcoholics and addicts include: the ''gorilla'' game of intimidation; the perpetual jokester or entertainer who has no serious problems; exaggerating depression; acting crazy; projecting an image of extreme fragility, weakness, inadequacy, helplessness, befuddlement, and indecisiveness; acting the part of the perfect patient; compulsively challenging all rules and authority in order to get discharged from treatment; acting as an assistant therapist for others' problems; totally absorbing one's self with music or sports while in treatment; and using sex to allay anxieties and/or get discharged from treatment. Both alcoholics and addicts play these games with varying degrees of awareness and unawareness.

In doing the literature search for this chapter, two articles[32,33] were discovered which indicated that several decades ago other clinicians were also stressing the similarities between alcoholics and addicts with respect to etiology, dynamics, defenses, and treatment. This discovery may cause one to suspect that the separating of these two troubled groups for the purpose of treatment was based more on socioeconomic and political grounds than on the weight of any scientific evidence.[34]

Finally, the therapeutic process of substance abuse rehabilitation, regardless of the substance(s) abused, is essentially the same: detoxification, psychobiosocial evaluation and history taking; restoration of physical health through proper diet, rest, and activities; addictive education; individual and/or group therapy designed to challenge defenses, facilitate catharsis, promote insight, instill hope, encourage spiritual renewal, and facilitate the learning of new modes of coping with personal needs and social demands; and providing for adequate after-care services (eg, social support systems such as AA and NA).

PERSONALITY AND PSYCHOPATHOLOGY SIMILARITIES AND DIFFERENCES

The Addictive Personality

Any examination of the similarities and differences in personality and psychopathology between alcoholics and addicts must also deal with the concept of the addictive personality. Basically, the term addictive personality has been used in the following ways:

1. To imply that there is a particular personality configuration that predisposes or causes an individual to become addicted. Some writers have implied an even greater degree of specificity, contending that certain personality traits predispose or cause a person to seek out and abuse certain chemicals but not others.
2. To imply that certain personality traits are observed only among addicted men and women or, at the very least, that these traits may be observed among addicted men and women in more virulent forms than among people not addicted to chemicals.

A review of the literature leads one to conclude that there is *no consistent support* for either meaning of the concept of an addictive personality. In fact, Landis[35] reached the same conclusion more than three decades ago when he stated that there was no typical alcoholic personality structure. Heavy drinkers, he noted, merely functioned differently than others. Typical of the basic inconsistency in the literature was Manson's[36] then contemporary counterclaim that he had found evidence for an addictive personality.

Mayberg,[37] after reviewing more than 150 references in connection with his research effort to examine how alcoholics and amphetamine abusers were similar or different, concluded, "...there appears to be no specific personality type that predisposes an individual to abuse alcohol, nor a personality type that predisposes amphetamine use" (p 23).

Miller's[38] study of alcoholism led to a similar conclusion: "One could conclude from this research that the average alcoholic is a passive, overactive, inhibited, acting-out, withdrawn, gregarious psychopath with a conscience, defending against poor defenses as a result of excessive and insufficient mothering" (p 657).

Thornburg's[39] search of the literature, too, did not uncover any consistent evidence in support of either an alcoholic personality or a drug-dependent personality. Her literature review and personal research study led her to conclude: "There does not seem to be a personality type or set of characteristics which predictably fit either addicts or alcoholics" (p 63). Her findings do indicate considerable within-group variations, suggesting there may be greater differences among alcoholics and among addicts then between alcoholics and addicts.

Regarding the second contention that certain personality traits are either found only among addicted men and women or that these traits are more exaggerated or intense when observed among the chemically addicted as compared to other groups, Sutherland et al,[40] Syme,[41] and

Lisansky[42] all concluded that no "hard" evidence existed to support the belief that alcoholics can be considered as different personalities than nonalcoholics or that there are any definite "signs" that can be used to differentiate them from other clinical or "normal" groups.

Nathan and Lansky[43] also question the validity of the first meaning of the term addictive personality, stating their position as:

> There do not appear to be characteristic personality patterns that differentiate drug abusers from nonabusers, there is clearly not a single route to alcohol or drug dependence, and alcoholics and nonalcoholics do not appear to metabolize ethanol in discreetly different ways at comparable levels of ingestion. Instead, the mechanism of dependence, the etiologic process, the personality structure of alcohol- and drug-dependent individuals, and their motivation for and response to treatment all depend on far more than intrapsychic or physiologic factors alone. To this end, the literature now supports a more sophisticated view, that of a complex individual system interacting with personal history and environmental factors to yield an addiction (p 714).

In Mayberg's[37] literature review, he found that the following personality traits had been identified by various writers and researchers as characteristic of alcoholics: high anxiety[44]; low frustration tolerance, alienation, and low self-concept[45]; psychopathic[46]; neurotic[47]; rebellious, impulsive, socially aggressive, gregarious, and not bound by social customs[48]; underlying dependency conflicts[49]; poor controls over impulsivity and aggressiveness[50]; and defiant of authority.[51]

Some of the personality traits reported in the drug abuse studies contained in Thornburg's[39] literature review were the following. Knight and Prout[52] described the addictive personality as psychopathic, impulsive, introverted, insecure, and shy. Boyd et al,[53] in a study of adolescent drug abusers, described their subjects as highly suspicious, especially of authority, but dependent and demanding, emotionally immature and guilt-ridden, overly sensitive to criticism and failure, with low tolerance for anxiety and frustration. Chein[54] also stressed low panic and frustration thresholds, profound distrust, and general depression and futility.

Mayberg,[37] who concentrated his review on amphetamine abusers, noted the following personality traits associated with that addiction. Low self-esteem, marked sensitivity to expressions of approval, or disapproval was reported by Wikler and Rasor.[55] McGrath,[56] however, disagreed with the notion that amphetamine abusers had low self-esteem, contending instead that they were hedonistic. Ellinwood[57] described the amphetamine abusers he studied as helpless, inept, and shameful, loners with no friends at school during their childhood.

Rosenberg[58] stressed the presence of conflict with parents and the lack of goals in life.

Even the most inexperienced of practitioners in the substance-abuse fields cannot help but note that the personality traits just listed are found frequently among *both* alcoholics and addicts and, for that matter, nearly any other group of people labeled as psychiatrically disturbed.

What is one to conclude, therefore, other than that there is no addictive personality? The concept is most likely to be employed by people who also subscribe to a stereotypic notion of *the* alcoholic or *the* addict. What should be evident to those working in the substance-abuse fields, however, is that there is great heterogeneity of personality and psychopathology to be observed among alcoholics[45,59,60,61–66] and drug-dependent men and women.[64,67–69]

On the basis of research findings and published clinical observations, it would seem far more valid to state that the differences *among* alcoholics and *among* drug addicts with respect to their personalities and mental health statuses are far greater than the differences that exist *between* these two categories of substance abusers. What does uniquely distinguish the addicted person would seem to be restricted to the person's *manner* of relating to various chemicals, the *circumstances* under which the person uses these chemicals, and the *psychobiosocial consequences* of such use.

Differential Reactions Associated with Personality

A related issue concerns the possibility that differential reactions to various chemicals, including placebos, may be mediated by personality variables. Lasagna et al,[70] for example, compared placebo-reactors to nonreactors using a projective test, the Rorschach. Reactors were described as being more anxious, self-centered, externalized, extroverted, and verbal than nonreactors.

Kornetsky and Humphries'[71] data also suggest a relationship between the intensity of response to psychoactive substances and personality. They used the Minnesota Multiphasic Personality Inventory (MMPI) to assess personality. Subjects in their study who responded most intensely to psychoactive medication was described as being more distressed and neurotic with respect to their MMPI scores (elevated *Hs, D, Hy,* and *Pt* scales).

Klerman et al,[72] using MMPI data, delineated two personality types, Type A and Type B, who reacted differently to sedatives and stimulants. Their Type A (elevated *Ma* and *Es* scales, low scores on the *D, Se,* and Manifest Anxiety scales) was described as action-oriented, extrapunitive, father-identified, athletic, practical, and had religious affiliations. These subjects were threatened by sedatives; these subjects reported that stimulants reduced anxiety. They also used alcohol considerably.

Their Type B subjects (elevated *D, Si,* and Manifest Anxiety scales, low scores on *Ma* and *Es*) were described as intropunitive, asthenic, mother-identified, and theoretically minded. These subjects reported finding stimulants threatening, while sedatives were perceived as anxiety reducing.

Heninger et al[73] reported that Type A and Type B subjects also reacted significantly differently to both phenothiazine and psychedelics. Frostad et al[74] discovered similar differential reactivities to diazepam.

Mayberg[37] summarized his review of this literature on differential reactions associated with personality as follows: "The experimental results suggest that response to any psychoactive drug is determined in part by the personality structure" (p 21). This conclusion seems justified, as long as one focuses on the phrase "in part," since the reaction (R) a person will experience to a substance is most likely mediated by the dynamic interaction of the person (P), including the person's physical and psychological makeup, the environment or situation surrounding such use (E), and the substance itself (S), thus $R = f(P \times E \times S)$.

Common Research Methodological Shortcomings Associated with Inconsistent Findings

To determine the extent to which the substance of abuse (alcohol *vs* another drug) accounts for personality and psychopathological similarities or differences, it is absolutely essential to remove or control for the influences of other variables, such as age, race, sex, and socioeconomic status. As basic as this rule of sound research may be, many research studies dealing with personality and psychopathological similarities and differences have failed to adhere to this rule.

The second most serious shortcoming noted in the literature relates to the classification of addictions. No uniform standards have been employed for judging whether or not someone is addicted to a particular substance (eg, many youths have been labeled "alcoholics" because they reported having gotten drunk on a weekend), or equating the intensity of addiction in terms of psychobiosocial consequences.

In addition, researchers tend to group all drug-dependent persons together (eg, heroin addicts, amphetamine abusers, barbiturate addicts) regardless of the substance(s) they had abused in the past or were abusing at the time the study was done. Moreover, various systems have been used for designating a particular substance of abuse as primary, secondary, etc in the case of multiple substance abuse. While these systems may satisfy the researchers' need for "operational definitions," they often seriously oversimplify the complex, dynamic interrelatedness and clinical significance of the individual's abuse of various substances.

With respect to psychopathology, *when* the data is collected (eg, during detoxification *vs* post-detoxification), *how* it is collected (eg, using a pathologically oriented instrument like the MMPI *vs* an instrument measuring "normal" personality traits, such as the PRF), *who* collects it (eg, a nonrecovered, white, mental health professional *vs* a black, recovered staff), the *circumstances* surrounding the data gathering (eg, whether the data are obtained voluntarily or not; the level of trust established), and the *anticipated consequences* of providing the data (eg, whether or not the testee believes the test results may influence any pending court actions). These factors have been highly variable from study to study, which no doubt accounts for much of the inconsistency noted in the literature.

Finally, the instrumentation used to assess personality and psychopathology have varied widely and, as any psychometrist well realizes, the *reliability* and *validity* of these instruments are *not* uniform. The most frequently used instrument for research purposes has been the Minnesota Multiphasic Personality Inventory (MMPI). The MMPI, unfortunately, yields a very narrow perspective of personality, since it essentially concerns itself with various forms of "psychopathology." This is one of the reasons we have learned very little about the "normal" personality dynamics of addicts and alcoholics.

Another instrument used for research purposes is the Rorschach, a projective test, the results of which are as much dependent on the testee's responses as on the testing circumstances and the relative skills and mental status of the examiner (psychometrists have been known to project their own unresolved problems on to their examinees from time to time).

Results of MMPI Studies

The intent and scope of this chapter do not permit an all-inclusive, in-depth review of the MMPI studies relating to similarities and differences between alcoholics and addicts. For convenience, the following sample of illustrative studies are grouped according to whether or not the researcher believed the data indicated greater degrees of differences than similarities between alcoholics and drug-dependent persons or vice versa.

MMPI studies indicating greater differences:

Stanton,[75] in an effort to determine whether or not a discriminating personality of criminal offenders could be developed, used MMPI to obtain scores on 100 white and 100 black prisoners. This sample included 35 white alcoholics and 20 black alcoholics, and 17 white addicts and 38 black addicts. Alcoholics scored significantly higher on the *Pd* scale than the narcotic addicts. His alcoholic prisoners also differed significantly from his nonaddicted prisoners.

Overall[76] reported that both alcoholics and narcotic addicts had elevated *Pd* scores; however, the alcoholics' profile had "substantially more" *D, Hy, Pt,* and *Sc* (alcoholics were pictured as neurotic, depressive, anxious, passive, and dependent), whereas the narcotic addicts' profile had higher *K* and *Ma* components (their profile type was characterized as antisocial, amoral, impulsive, irritable, hostile, and psychopathic). Unfortunately, Overall did not control for the effects of age or other variables that may have influenced these results, eg, race.

McLachlan[64] demonstrated that Overall's[76] original discriminant function (DF) technique discriminated alcoholics from narcotic addicts with 65% accuracy, which is less than that originally reported by Overall who reported 85% accuracy. McLachlan raised the accuracy level to 80% by screening out "normal" MMPIs. The DF technique was affected by sex and age.

Lachar et al[77] compared the MMPI scores of 65 alcoholics, 48 heroin addicts, and 52 polydrug abusers with those of a demographically matched group of "psychiatric controls." All of the men studied had volunteered for treatment. Whereas no significant differences were noted between alcoholics and heroin addicts, the polydrug sample did describe themselves as experiencing significantly more psychiatric symptomatology, although this finding must be tempered by the fact that the polydrug abusers were "less naively defensive" in responding to the MMPI items. Unfortunately, race and age differences among these three groups of substance abusers were not controlled for in the analysis even though 83% of the polydrug abusers were white *vs* 49% white for alcoholics and 29% white for heroin addicts, and the mean age for the alcoholics was 42.17 *vs* 25.21 for the polydrug sample.

MMPI studies indicating greater similarities:

Sutherland et al[40] reviewed 37 research reports on personality characteristics of "chronic alcoholics." They concluded that "no satisfactory evidence has been discovered that justifies a conclusion that persons of one type are more likely to become alcoholics than persons of another type." They agreed with Wexberg[78] who earlier had asserted, "There is no alcoholic personality prior to alcoholism."

Hill[79] emphasized deficient social controls, insufficient "counter-anxiety," and inadequate reinforcement from daily activities common to *both* alcoholics and opiate addicts.

Hill et al,[80] in an MMPI study of institutionalized groups of alcoholics, narcotic addicts, and criminals, noted remarkably similar MMPI psychographs or profiles for all three groups. They reported small, statistically significant, "but nondiagnostic differences" on three-factor, analytically derived factors.

Haertzen et al[81] reported "small but significant" differences between alcoholics and addicts (0.7 of a standard deviation when only white subjects were compared) with respect to personal pathology and personality questions.

Rardin et al[82] compared opiate-, amphetamine-, and alcohol-abusing soldiers seeking treatment. They noted "striking" similarities across all three groups of substance abusers; the MMPI profiles were very similar in configuration.

Black and Heald[83] compared 40 male alcoholics and 50 male illicit drug abusers from a military drug and alcohol rehabilitation program. They reported finding similar MMPI profile configurations for the two groups. They concluded that, whereas the drug abusers appeared somewhat more psychiatrically disturbed than the alcoholics, the two groups "cannot be differentiated to any significant degree with regard to current personality functioning."

Mayberg[37] compared 159 white drug addicts with 186 white alcoholics residing in a VA hospital; the alcoholics were subdivided into three subgroups: 1) "young alcoholics" (under age 30); 2) "old young alcoholics" (their drinking problems started early); and 3) "old alcoholics" (their drinking problems developed later in life). Of the drug addicts, 95% were under 30. Mayberg reported significant differences among his alcohol groups; he also noted very few differences between the young alcoholics and the amphetamine and polydrug user. "The most obvious difference was the Young Alcoholic's need to assert his masculinity and the drug user's apparent paranoia. With these two exceptions, the (MMPI) profiles were quite similar" (p 156).

Thornburg[39] studied 40 male alcoholics and 40 male drug addicts who had completed a VA chemical-dependency program. All of the men were between 19 and 33 years old; all but one were Caucasian. The alcoholics were much older than her drug-dependent sample. While reporting some differences between alcoholics and addicts, both on the pre-treatment (high *Mf, F*) and post-treatment (*Mf, F, ?,* and *Ma*) MMPI scores, she concluded that "the present study supplies no evidence to support the substantial personality differences revealed in other studies" (p 283). It should be noted that the significant difference she reported for the *Ma* scale could just as well be accounted for by age as by substance abuse.

MMPI Conclusions

What does one learn from the MMPI studies comparing alcoholics and addicts? First and foremost, since *both* similarities and differences have been reported, the findings are obviously inconsistent. Even where significant differences did emerge, the *same* differences between the two groups were not consistently reported (eg, sometimes alcoholics had higher *Pd* scores, while addicts scored higher in other studies). Second, most of the MMPI studies failed to adequately control for the influence of such important intervening variables as age, race, sex, and socioeconomic factors. Third, some of the researchers reporting statistically significant differences questioned the clinical utility of these differences, since the magnitudes of the differences reported were small. Fourth, even where some MMPI scales yielded significant differences, the majority of MMPI scales did not. Fifth, in some studies reporting small, but statistically significant differences,[82] the MMPI psychographs (ie, the rise and fall of the individual scale scores) were very similar in their configuration.

One might well conclude from these studies that greater similarities exist than differences, since the reported differences are inconsistent in their nature, often small in magnitude or restricted to a minority of the MMPI scales. These differences could just as well be accounted for by intervening variables (eg, age, race, sex), which very often were poorly controlled for as by the substance of abuse. Studies, such as those done by Aaronson,[84] Calden and Hokanson,[85] McGinnis and Ryan,[86] and Gynther,[87] very clearly have demonstrated that age, race, and sex have affected MMPI scale scores.

Special Scales Developed from the MMPI

A number of researchers have attempted to develop special scales to diagnose and/or predict susceptibility to narcotic addiction[88] and alcoholism[89-94] based on item analyses of the MMPI. Of these, the MacAndrew (AMac) scale has fared best in the research literature.

The AMac scale consists of 49 items from the MMPI. On these items, alcoholics typically describe themselves as bold, out-going, and social, having few conflicts or problems; they do, however, admit to having had problems in school and resent authority; they seem drawn to religion, faith, and inspiration; and they claim to be better off than neurotics but report more bodily complaints.

Whisler and Cantor[95] reported that the AMac scale discriminated between alcoholic and nonalcoholic patients in a "chronic and institutionalized population." Using MacAndrew's[92] cutoff score of 24, they reported 55% accuracy in classifying the two groups, with 7.9% "false negatives" and 37.1% "false positives." Accuracy was improved to 61.5% by using 28 as a cutoff point.

Rhodes's[96] study demonstrated that the AMac discriminated between alcoholics and nonalcoholics in an outpatient psychiatric setting.

Rohan et al[97] examined the MMPI scores of alcoholic subjects before and after treatment. They observed that whereas their subjects' clinical MMPI scales generally improved with treatment, the AMac scale scores remained essentially unchanged. This suggested to them that the AMac scale might be measuring a relatively stable personality trait.

Vega[98] reported that the AMac scale correctly discriminated alcoholics from "normals" and psychiatric patients, although "false positives" were noted among the normal group. Hoffman et al[99] did a longitudinal study in which they were able to discriminate *pre*-alcoholic college freshmen from their peers using the AMac scale.

Additional studies by Chang et al[100] and Jacobson [101] have also attested to the AMac scale's ability to distinguish between alcoholics and nonalcoholics.

In the only study attempting to differentiate alcoholics from heroin addicts and from non-chemically dependent people, Kranitz[102] found that the AMac scale did *not* differentiate between alcoholics and heroin addicts. The scale did, however, differentiate both inpatient alcoholics and heroin addicts from inpatient and outpatient nonalcoholics. This led Kranitz to conclude that MacAndrew had not developed an alcoholism scale per se.

If the AMac scale could successfully differentiate alcoholics from nonalcoholics, concurrent validity could be claimed for the scale. In the same vein, if the AMac scale could successfully predict alcoholism, predictive validity could be argued for the scale. Both findings have been reported, and these results would seem to suggest that there may indeed be a personality trait unique to alcoholics.

However, several notes of caution need to be sounded. First, several studies reported fairly high levels of misclassification.[95] Second, the scale did not differentiate alcoholics from addicts.[102] Third, and perhaps most important, MacAndrew has *not* removed all the MMPI items that relate directly to alcohol abuse. While MacAndrew[92] did eliminate MMPI items 215, "I have used alcohol excessively"—*true,* and 460, "I have used alcohol moderately (or not at all)"—*false,* he left in the following MMPI items:

156 "I have had periods in which I carried on activities without knowing later what I had been doing"—*true*
186 "I frequently notice my hand shakes when I try to do something—*true*
130 "I have never vomited blood or coughed up blood"—*false*
294 "I have never been in trouble with the law"—*false*

By leaving in MMPI items such as the preceding, MacAndrew confounds personality with styles and consequences of abusive drinking and makes it impossible to properly ascertain whether there is or is not a stable, underlying personality trait unique to alcoholics. On the basis of the three preceding notes of caution, it cannot be said that a stable personality trait, unique to alcoholics, has yet been established using the AMac scale.

Non-MMPI Studies Comparing Alcoholics and Addicts

Surrey[103] compared the performance of a matched group of heroin addicts or "heroinics," alcoholics, and controls using Witkin's[104] Rod and Frame Test (RFT). To perform well on the RFT, a subject must ignore external distracting clues and rely on internal stimuli in making perceptual judgments regarding the relative position of a rod vis-a-vis a surrounding frame. Surrey compared these three groups under two conditions, nonstressed and stressed (white noise was the source of stress).

Surrey reported heroinics as being more field-dependent (reliant on external clues) than alcoholics; both addicted groups in turn were reported to be significantly more dependent than her control group. She also reported that heroinics and alcoholics were significantly more vulnerable to stress than her control group.

Ballner[105] sought to extend these findings using different measures of dependency, namely the Thematic Apperception Test (TAT) and the Holtzman Inkblot Test (HIT) projective tests. He also used a different procedure for stressing his subjects. Ballner's heroinics were observed to be more anxious and dependent than his matched group of alcoholics and controls. His data, however, did not indicate any consistent increase in anxiety and dependency for either his heroinics or alcoholics as a result of being stressed.

Ballner[105] also cited a study by Rauchfleisch[106] using the Rorschach comparing matched groups of heroinics, alcoholics, and controls. Although no data are given, Rauchfleisch contended that his heroinics were more labile and oral-fixated than the other two groups.

Brien et al[107] compared the performance of 25 heroin addicts, 25 alcoholics, 25 methamphetamine abusers, and 25 individuals with mixed drug dependencies using the 16PF. Their results indicated that none of the four groups differed greatly from the norm, leading the researchers to ponder whether or not drug users were really much different from the general population. An alternative explanation offered was that the 16PF may not have been sensitive enough to discern differences. All groups but the alcoholics scored low on Factor C (a purported measure of ego strength). The methamphetamine abusers had the most deviant scores of the four groups studied.

During the 1970s a series of studies emerged concerning the relationship between a personality characteristic known as "sensation-seeking" and substance abuse.[108-112] While these researchers did not directly address the issue of similarities and differences between alcoholics and addicts, their findings are of interest to us.

Zuckerman et al,[109] while reporting that sensation-seeking was related to their subjects' sexual and drug-taking experiences, noted that the proportionate increase in drinking and marijuana use was *greatest* for *low* sensation-seeking males. Earlier, Zuckerman et al[108] had reported that greater drug usage, alcohol intake, smoking marijuana, and varieties of sexual experience were associated with high sensation-seeking in both sexes.

Kaestner et al[110] reported finding an association between sensation-seeking and anxiety, and the number of different drugs used by whites (but not nonwhites). Schwarz et al[111] reported a strong positive correlation between sensation-seeking and drinking. Sutker et al[112] also reported an association between sensation-seeking and substance-abuse patterns.

The sensation-seeking studies would seem to indicate that high sensation-seeking is associated with *both* increased drinking *and* the use of a greater number of drug categories, although the relationship is not uniform across race and sex.

Two Eagleville Studies of Similarities and Differences of Personality and Psychopathology

As previously stated, any effort to examine similarities and differences between alcoholics and addicts must attempt to eliminate the effects of other variables that might influence measures of personality and psychopathology. To this end, two studies were initiated at Eagleville Hospital and Rehabilitation Center (EHRC) by Carroll et al.[34,113] In these studies, statistical procedures† were employed for partialling out the relative contribution of age and race.

The subjects included 178 male alcoholics and 156 male addicts; 170 were black and 164 were white. The addicts were very nearly all heroin abusers, although some of the men had also abused other drugs. The diagnosis of alcoholic or drug-dependency had been determined at the time of admission by extant hospital policies and procedures.

To examine the similarities and differences between alcoholics and addicts with respect to personality, two standard psychological instruments were used: the Personality Research Form (PRF by Jackson[115]), and the clinical and research form of the Tennessee Self-Concept Scale (TSCS by Fitts[116]).

Results for Personality Comparisons

The data were examined from two perspectives, one that considered *only* the classification of alcoholic *vs* addict, and another that considered the impact of this classification when the effects of age and race had been partialled out through a multivariate statistical analysis of the data. Since only males were included in this study, the impact of biological sex differences was not a problem to contend with. In a similar vein, nearly everyone who is admitted to EHRC for treatment is on Medical Assistance; thus, at the time of admission, the patient population was very homogeneous with respect to socioeconomic status—they were, with very few exceptions, quite poor.

Considering *only* the addiction classification, the following results were obtained* (see Figure 4-1).

Alcoholics Significantly Higher	Addicts Significantly Higher
Abasement*	Aggression**
Achievement**	Autonomy**
Cognitive Structure**	Change**
Harm-avoidance**	Dominance*
Order**	Exhibition**
	Impulsivity**
	Play**

*$p \leqslant .05$
**$p \leqslant .01$

†A three-way analysis of variance and covariance using an SPSS model for factoral design with unequal cell frequency, subprogram ANOVA.[114]

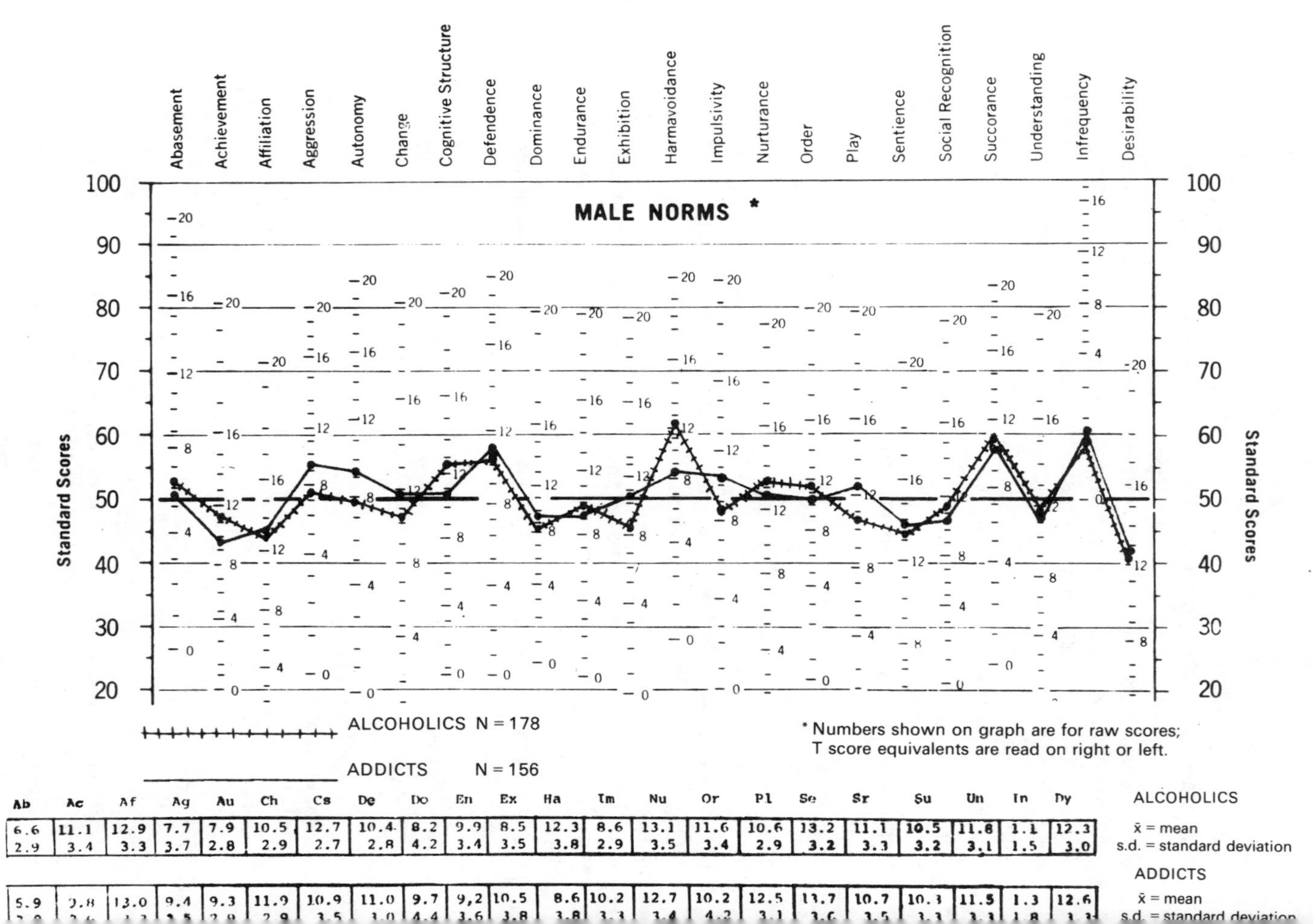

	Ab	Ac	Af	Ag	Au	Ch	Cs	De	Do	En	Ex	Ha	Im	Nu	Or	Pl	Se	Sr	Su	Un	In	Dy
ALCOHOLICS x̄	6.6	11.1	12.9	7.7	7.9	10.5	12.7	10.4	8.2	9.9	8.5	12.3	8.6	13.1	11.6	10.6	13.2	11.1	10.5	11.8	1.1	12.3
ALCOHOLICS s.d.	2.9	3.4	3.3	3.7	2.8	2.9	2.7	2.8	4.2	3.4	3.5	3.8	2.9	3.5	3.4	2.9	3.2	3.3	3.2	3.1	1.5	3.0
ADDICTS x̄	5.9	9.8	13.0	9.4	9.3	11.9	10.9	11.0	9.7	9.2	10.5	8.6	10.2	12.7	10.2	12.5	13.7	10.7	10.3	11.5	1.3	12.6
ADDICTS s.d.	[illegible]	[illegible]	[illegible]	3.5	2.9	2.9	3.5	3.0	4.4	3.6	3.8	3.8	3.3	3.4	4.2	3.1	3.6	3.5	3.3	3.3	1.8	3.3

Figure 4-1

These differences could be attributed to the extent to which the two groups accept or reject prevailing middle-class societal norms and whether or not they adopt a relatively active or passive coping stance toward society.

Alcoholics, as reflected by their higher *Achievement* scale scores, seem to adhere more to the middle-class American dream or struggle and achievement, including the modal values underlying this perspective. Their higher *Abasement* scale score reflects their greater tendency to blame themselves for their failure to achieve these goals.

The higher *Cognitive Structure* and *Order* scale scores of the alcoholics indicate a stronger need for structure and consistency in their environment, to know better and more clearly the "rules of society's game."

The alcoholics' higher *Harm-avoidance* scale score, together with their higher need for *Order* and *Cognitive Structure,* reflects a greater need to cope with internal distress and external stress by seeking maximal security in the order and regularity that their environment provides.

Alcoholic men thus seem to be more willing than addicted men to employ a passive, dependent strategy in order to obtain order and regularity in their life experiences. Their hope for a better future, therefore, seems destined to be determined by external forces over which they, the alcoholics, exercise little control.

Addicts, by contrast, appear to be far more alienated from the middle-class societal norm of striving to achieve traditional goals through traditional means. They also seem less committed to the societal status quo and more willing to experiment with new and different experiences as reflected by their higher *Change* scale score. Moreover, when addicts encounter failure, unlike their alcoholic counterparts, tend to reject personal responsibility for such failure.

Addicts also are more inclined to attempt to assert themselves in gaining control over their environment, as evidenced by their higher *Aggression* and *Dominance* scale scores. They appear more uninhibited, rash, pleasure-seeking, and oriented to the here and now, as indicated by their higher *Impulsivity* and *Play* scale scores.

Rather than turn anger and aggression inwardly, as the alcoholics apparently did, addicts are more inclined to turn these feelings and actions against others and society. Unlike the alcoholics, the addicts seemed more inclined to go their own way and seek a better life through their own actions as reflected by their higher *Autonomy* scale score.

These interpretations did not seem exceptional or unusual; in fact, they seemed rather stereotypic. We suspected, however, that these conclusions warranted more careful scrutiny; therefore, a second analysis of these data was done. This analysis made it possible to evaluate the main effects of addiction (alcoholic *vs* addict), while partialling out the effects of age and race (sex and socioeconomic status, as previously noted, had been "controlled for" by virtue of the patient population under study). When this was done, the results changed dramatically.

Addiction, as a main effect independent of age and race, yielded only *one PRF* scale where alcoholics were significantly different from addicts. The other 19 *PRF* scales produced *no* statistically significant differences. A visual inspection of Figure 4-1, which depicts a very similar *PRF* profile for the two groups, would seem to lend visual support to this finding.

The statistically significant difference occurred on the *Desirability* scale, which is actually a response set indicator rather than a personality need scale per se. This finding stands in sharp contrast to the findings previously described for the *t* test analysis, which yielded significant differences on 12 of the 20 *PRF* scales. In addition, finding only one out of 20 comparisons statistically significant raises the statistical probability that this result could be due to chance.

Addicts scored significantly higher than alcoholics on the *Desirability* scale. This finding is in accord with the general trend of the literature in the addiction problems field.[26,117] Addicts generally are reported as being more manipulative, more cunning, and better able to put on a good front than alcoholics. All of these traits may be considered as essential ingredients for survival on the street in the procurement and use of illicit substances.

The only significant interaction effect was for addiction and age. That was for the *PRF Abasement* scale. The data indicated that young alcoholics scored significantly higher than young addicts on this scale. This was most clear for blacks, even though no significant interaction effect for race was obtained. The interaction analyses unfortunately were attenuated due to the dearth of young, white alcoholics. The reader should be cautioned, therefore, that the findings might have been more sound had the number of cases been more equally distributed across all the variables.

Subsequent to this study, a second independent study was conducted at EHRC by Cohen et al.[118] These researchers compared the *PRF* test results of a second sample of 109 alcoholics and 136 heroin addicts in treatment at Eagleville.

Their findings were essentially the same as those obtained several years earlier, namely that when age and sex differences between the two groups were partialled out, alcoholics and addicts showed far more similarities than differences in their personality structure as measured by the *PRF.* On the only two scales (the *PRF* has 22 scales) that yielded statistically significant differences, alcoholics scored higher than heroin addicts on the *Harm-avoidance* and *Affiliation* scales. This would suggest a greater level of fear and anxiety on the part of the alcoholic patients and a greater need to be with others. By contrast, the addicts appeared to acknowledge or "own" less fear and anxiety and express a greater sense of alienation in that they claimed to need people to a lesser degree than alcoholics.

Self-Concept Data

When the *TSCS* data for the same subjects used in the Carroll et al[34] *PRF* study were subjected to a simple bivariate analysis, alcoholic *vs* addict, the following results were obtained (see Figure 4-2).

Alcoholics Significantly Higher	Addicts Significantly Higher
T/F Ratio**	Total Positive**
Number of 4's ("Mostly True")*	Self Acceptance**
	Physical Self**
	Personal Self**
	Social Self**

*$p \leqslant .05$
**$p \leqslant .01$

These results show a stronger tendency on the part of alcoholics to agree with items regardless of their content (higher *T/F Ratio* and *Number of 4's*). Generally this is believed to be indicative of an acquiescence response set and lends support to the popular notion that alcoholics are more passive and compliant than addicts.

The "higher" scores achieved by addicts on the *TSCS* phenomenological scales *(Total Positive, Self Acceptance, Physical Self, Personal Self,* and *Social Self)* indicate that either the addicts being studied had not sunk as low as the sample of alcoholics in self-esteem, or they were not admitting it. Both groups evidenced considerably less self-esteem than the general population. For example, using the *Total Positive* scale as the single best measure of self-esteem on the *TSCS,*[116] we would expect 90% of the general population to score higher on this scale than did the two groups of substance abusers.

The profile configuration of both groups was also strikingly similar. Both the alcoholics and addicts, for example, evidenced a characteristic dip on the *Moral-Ethical* and *Family Self TSCS* scales. These depressed scores reflect a *common* underlying need of both alcoholics *and* addicts to work through their unresolved guilt, espe-

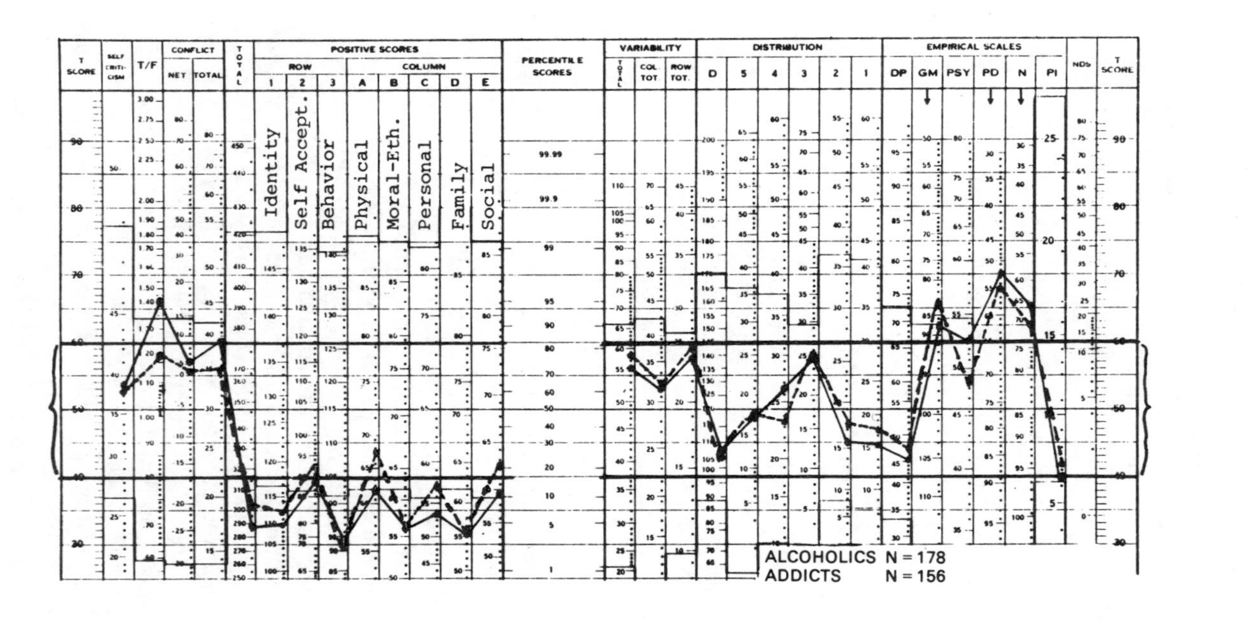

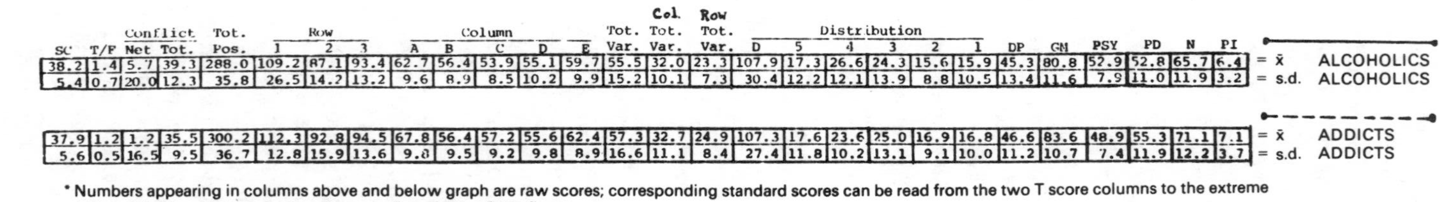

SC	T/F	Conflict Net	Conflict Tot.	Tot. Pos.	Row 1	Row 2	Row 3	Column A	Column B	Column C	Column D	Column E	Tot. Var.	Col. Tot. Var.	Row Tot. Var.	Distribution D	Distribution 5	Distribution 4	Distribution 3	Distribution 2	Distribution 1	DP	GM	PSY	PD	N	PI	
38.2	1.4	5.7	39.3	288.0	109.2	87.1	93.4	62.7	56.4	53.9	55.1	59.7	55.5	32.0	23.3	107.9	17.3	26.6	24.3	15.6	15.9	45.3	80.8	52.9	52.8	65.7	6.4	= x̄ ALCOHOLICS
5.4	0.7	20.0	12.3	35.8	26.5	14.2	13.2	9.6	8.9	8.5	10.2	9.9	15.2	10.1	7.3	30.4	12.2	12.1	13.9	8.8	10.5	13.4	11.6	7.9	11.0	11.9	3.2	= s.d. ALCOHOLICS
37.9	1.2	1.2	35.5	300.2	112.3	92.8	94.5	67.8	56.4	57.2	55.6	62.4	57.3	32.7	24.9	107.3	17.6	23.6	25.0	16.9	16.8	46.6	83.6	48.9	55.3	71.1	7.1	= x̄ ADDICTS
5.6	0.5	16.5	9.5	36.7	12.8	15.9	13.6	9.8	9.5	9.2	9.8	8.9	16.6	11.1	8.4	27.4	11.8	10.2	13.1	9.1	10.0	11.2	10.7	7.4	11.9	12.2	3.7	= s.d. ADDICTS

* Numbers appearing in columns above and below graph are raw scores; corresponding standard scores can be read from the two T score columns to the extreme left and right or the percentile column in the middle of the figure.

Figure 4-2

cially guilt associated with past misdeeds and failures involving family members.

When, as before, statistical controls* for the effects of age and race were exercised (since these were the same subjects as in the *PRF* study, we had, in effect, "controlled" for the effects of sex and socioeconomic status), only the *T/F Ratio* scale continued to yield a significant difference between the two groups of substance abusers. Figure 4-2 indicates that the two groups produced very similar *TSCS* profiles, which again seems to support this finding.

While we can place a higher degree of confidence in this difference, ie, alcoholics manifesting a greater degree of acquiescence than addicts, the reader must be cautioned to note that when the results for 28 *TSCS* scales were analyzed,† only three‡ continued to yield significant differences between the two groups of substance abusers in the multivariate analysis. This finding clearly points to greater similarities than differences between the two groups.

Psychopathology

When one considers the psychobiosocial consequences of an addictive lifestyle, especially the psychic cost, the differences between alcoholics and addicts with respect to psychopathology again appear to be far less significant than their similarities.

To name but a few of the symptoms of psychopathology observed among *both* of these groups, consider the following: acute and chronic forms of impaired cognitive functioning; high levels of anxiety, restlessness, and tension; numerous fears; high levels of distrust of others; loneliness and despair; feelings of helplessness and inadequacy; sexual maladjustments; inability to express and/or accept love and friendship; uncontrollable outbursts of anger and violence; and periodic, unpredictable psychotic episodes, including hallucinations, delusions, and insomnia and depression.

As previously noted, the within-group variance (eg, how addicts differ among themselves) is far greater than the between-group variance, ie, how addicts differ from alcoholics with respect to their psychopathology. This generalization assumes that the issue of psychopathology and substance abuse is being considered apart from differences in psychopathology associated with age, race, sex, and socioeconomic status.

---

*A multivariate analysis of variance and covariance.

†The NDS scale was not included in this analysis.

‡The two other significant findings for the *Psychosis* and *Personality Disorder* scales will be discussed in the section on psychopathology.

Returning to the previously cited *TSCS* study[34] comparing alcoholics and addicts, when one considers *only* the diagnostic distinction between being an alcoholic *vs* being an addict,* we observed the following results on the *TSCS* empirical scales measuring psychopathology (see Figure 4-2).

Alcoholics Significantly Higher	Addicts Significantly Higher
Net Conflict**	None
Total Conflict**	
General Maladjustment*	
Psychosis**	
Personality Disorder*	
Neurosis**	

*$p \leqslant .05$
**$p \leqslant .01$
Inverted scales, lower raw scores indicate greater psychopathology.

The higher scores on the *TSCS* empirical scales indicated that the alcoholics manifested a higher level of general emotional distress and maladjustment than did the addicts (higher *General Maladjustment*), including greater confusion, conflict, fear, and anxiety (higher *Neurosis*); as well as poorer reality testing, greater emotional lability, and depression (higher *Psychosis*); and more personality weaknesses and defects (higher *Personality Disorder*).

When these same scales were analyzed in a manner that partialled out the effects of age and race, however, the picture changed considerably. Only the *Psychosis* and *Personality Disorder* scales continued to yield significant differences, with the alcoholics scoring higher as before. Again, the similarity in *TSCS* profiles depicted in Figure 4-2 would seem to support this finding. We can thus place somewhat more confidence in the findings that indicate higher levels of acknowledged emotional distress and poorer personality integration and functioning for alcoholics.

CONCLUSIONS

It is obvious that similarities and differences with respect to personality and psychopathology do exist between those who abuse alcohol and those who abuse other drugs. The challenge to those seeking to understand, treat, and prevent the abuse of substances,

*Again a simple *t* test was used for this analysis.

however, is to discern the basis for these similarities and differences and their clinical relevance. The evidence to date would seem to suggest that most of the differences observed are more a reflection of *age, race, sex, socioeconomic status,* and the use of an *illegal or legal substance* of abuse rather than any *intrapsychic dynamics.*

The studies that the author has conducted and reviewed, as well as his own clinical experience, lead him to conclude that the similarities in personality and psychopathology between alcohol abusers and drug abusers have greater clinical and social significance than do their differences. In addition, the importance and relevance of the distinction between those who abuse alcohol and those who abuse other drugs seem to be lessening as more and more substance abusers engage in multiple substance abuse. Overwhelmingly, multiple substance abuse involves the abuse of *both* alcohol and other drugs.

The common core that underlies *all* forms of substance abuse and addiction is a negative self-concept. With rare exceptions, substance abusers in treatment present with self-concepts replete with a sense of profound failure, weakness, inadequacy, guilt, shame, loneliness, self-doubt, confusion, and despair. Regardless of which substance(s) the substance abuser may have abused, the primary goal of treatment and rehabilitation is to restore self-respect or, for many of our patients, assist them for the first time ever in their lives to achieve a personal sense of self-worth and dignity.

The process by which this goal of achieving self-respect and personal dignity is to be pursued is "content-free," that is, it is generic rather than substance-specific or problem-specific.

REFERENCES

1. Rosenthal R: *Experimenter Effects in Behavioral Research.* New York, Appleton-Century Crofts, 1966.
2. Grier WH, Cobbs PM: *Black Rage.* New York, Basic Books, 1968.
3. Rosenthal R, Jacobson L: *Pygmalion in the Classroom: Teacher Expectation and Pupil's Intellectual Development.* New York, Holt, Rinehart, and Winston, 1968.
4. Temerlin MK: Suggestive effects in psychiatric diagnosis. *J Nerv Ment Dis* 147:349–359, 1968.
5. Weisstein N: Psychology constructs the female, in Gornick V, Moran BK (eds): *Women in Sexist Society: Studies in Power and Powerlessness.* New York, Basic Books, 1971, pp 133–146.
6. Chesler P: *Women and Madness.* Garden City, NY, Doubleday, 1972.
7. Abramowitz SI, Abramowitz CV, Jackson C, et al: The politics of clinical judgment: What nonliberal examiners infer about women who do not stifle themselves. *J Consult Clin Psychol* 41:385–391, 1973.

8. Rosenhan DL: On being sane in insane places. *Science* 179:250–258, 1973.

9. Aslin AL: Feminist and community mental health center psychotherapists' expectations of mental health for women. *Sex Roles* 3:537–544, 1977.

10. Freudenberger HL: The gay addict in a drug and alcohol abuse therapeutic community. *The Addiction Therapist* 2:23–30, 1977.

11. Morin SF: Heterosexual bias in psychological research on lesbianism and male homosexuality. *American Psychologist* 32:629–637, 1977.

12. Braginsky DD: The mentally ill, the alcoholic, the drug addicts: Misfits all, in Ottenberg DJ, Carpey EL (eds): *Proceedings of the 7th Annual Eagleville Conference.* Rockville, Md, Alcohol, Drug Abuse, and Mental Health Administration, 1974, pp 105–112.

13. Sagarin E: The high personal cost of wearing a label. *Psychology Today* 9:25–26, 30–31, 1976.

14. Carroll JFX: "Mental Illness and Disease": Outmoded concepts in alcohol and drug rehabilitation. *Comm Ment Health J* 11:418–429, 1975.

15. Carroll JFX: Mental illness and addiction: Perspectives which overemphasize differences and undervalue commonalities. *Contemp Drug Problems* 7:227–231, 1978.

16. Huberty DJ: The addict and the alcoholic in treatment. Some comparisons. *J Drug Issues* 3:341–347, 1973.

17. Robbins PR, Nugent JF: Perceived consequences of addiction: A comparison between alcoholics and heroin-addicted patients. *J Clin Psychol* 31:367–369, 1975.

18. Driscoll GZ: *Comparative study of drug dependent and alcoholic men at Eagleville Hospital and Rehabilitation Center.* Paper presented at the Alcohol Drug Problems Association Annual Conference, Hartford, Conn, September 1971.

19. Driscoll GZ, Barr HL: *Comparative study of drug dependent and alcoholic women.* Paper presented at the Alcohol Drug Problems Association Annual Conference, Atlanta, September 1972.

20. Ferneau EW: The drug abuser and the alcoholic: Some similarities. *Br J Addict* 66:71–75, 1971.

21. Belter EW: *Booze, hamburger, and noodles—Combining alcohol and other drug programming.* Paper presented at the North American Congress on Alcohol and Drug Problems, San Francisco, December 1974.

22. Cohen A: *Alcohol and heroin—Structural comparison of reasons for use between drug addicts and alcoholics.* Paper presented at the 6th Annual Scientific Meeting of the National Council on Alcoholism, Milwaukee, April 1975.

23. Chafetz M: Should alcohol and other drug programs be combined? Yes! *Wisconsin Alcohol and Other Drug Association Newsletter,* September 1976.

24. Carroll JFX, Malloy TE: Combined treatment of alcohol- and drug-dependent persons: A literature review and evaluation. *Am J Drug Alcohol Abuse* 4:343–364, 1977.

25. Neumann CP, Tamerin JS: The treatment of adult alcoholics and teenage drug addicts in one hospital: A comparison and critical appraisal of factors related to outcome. *Q J Stud Alcohol* 32:82–93, 1971.

26. Poze RS: Heroin addicts in a community mental health inpatient unit. *Am J Psychiatry* 129:120–124, 1972.

27. Fiedler FE: A comparison of psychoanalytic, nondirective, and Adlerian therapeutic relationships. *J Consult Psychol* 14:436–445, 1950.

28. Egan G: *The Skilled Helper.* Monterey, Calif, Brooks/Cole Publishing Co, 1975.

29. Carroll JFX, Malloy TE, Hannigan PC, et al: The meaning and evolution of the term "multiple substance abuse." *Contemp Drug Problems* 6:101–133, 1977.

30. Gerston A, Cohen MJ, Stimmel B: Alcoholism, heroin dependency, and methadone maintenance: Alternatives and aids to conventional methods of therapy. *Am J Drug Alcohol Abuse* 4:517–531, 1977.

31. Kaufman E: Polydrug abuse or multidrug misuse: It's here to stay. *Br J Addict* 45:462–468, 1977.

32. Gray MG, Moore M: A comparison of alcoholism and drug addiction with particular reference to the underlying psychopathological factors. *J Criminal Psychopath* 4:151–161, 1942–1943.

33. Gerard DL: Intoxication and addiction: Psychiatric observations on alcoholism and opiate drug addiction. *Q J Stud Alcohol* 16:681–699, 1955.

34. Carroll JFX, Klein MI, Santo Y: A comparison of the similarities and differences in the self-concept of male alcoholics and addicts. *J Consult Clin Psychol* 46:575–576, 1978.

35. Landis C: Theories of the alcoholic personality: Lecture 11. In Yale University, Laboratory of Applied Physiology, School of Alcohol Studies. *Alcohol, science, and society: Twenty-nine lectures with discussions, as given at the Yale Summer School of Alcohol Studies.* New Haven, Conn, *Q J Stud Alcohol* 129–142, 1945.

36. Manson MP: A psychometric differentiation of alcoholics from non-alcoholics. *Q J Stud Alcohol* 9:175–206, 1948.

37. Mayberg SW: *Alcohol/amphetamines: Use of the MMPI and life history data to determine factors influencing selection in a chemically dependent population,* doctoral dissertation. University of Minnesota, 1975. (Xerox University Microfilms, Ann Arbor, Mich, 75-27, 174)

38. Miller WR: Alcoholism scales and objective assessment methods: A review. *Psychol Bull* 83:649–674, 1976.

39. Thornburg S: *An examination of the single versus dual treatment controversy in chemical dependency: Differences between alcoholics and drug addicts,* doctoral dissertation. University of Minnesota, 1977. (Xerox University Microfilms, Ann Arbor, Mich, 77-26, 169)

40. Sutherland EH, Schroeder AM, Tordella AB: Personality traits and the alcoholic: A critique of existing studies. *Q J Stud Alcohol* 11:548–561, 1950.

41. Syme L: Personality characteristics of the alcoholic. *Q J Stud Alcohol* 18:288–302, 1957.

42. Lisansky ES: Clinical research in alcoholism and the use of psychological tests: A reevaluation, in Fox R (ed): *Alcoholism: Behavioral Research, and Therapeutic Approaches.* New York, Springer, 1967.

43. Nathan PE, Lansky D: Common methodological problems in research on the addictions. *J Consult Clin Psychol* 46:713–726, 1978.

44. Hobson GN: Anxiety and the alcoholic as measured by eye-blink conditioning. *Q J Stud Alcohol* 32:976–981, 1971.

45. Mogar RE, Wilson WM, Helen ST: Personality subtypes of male and female alcoholic patients. *Int J Addict* 5:99–113, 1970.

46. Glatt MM, Hills DR: Alcohol abuse and alcoholism in the young. *Br J Addict* 63:183–191, 1968.

47. Rosenberg CM: Determinants of psychiatric illness in young people. *Br J Psychiatry* 115:907–915, 1969a.

48. Loper RG, Kammeier ML, Hoffman H: MMPI characteristics of college freshmen males who later became alcoholics. *J Abnorm Psychol* 82:159–162, 1973.

49. Robins LN, Bates WM, O'Neal P: Adult drinking patterns of former problem children, in Pittman DJ, Snyder CR (eds): *Society, Culture, and Drinking Patterns.* New York, Wiley, 1962, pp 395–412.

50. Lisansky-Gomberg ES: Etiology of alcoholism. *J Consult Clin Psychol* 32:18–20, 1968.

51. Jones MC: Personality correlates and antecedents of drinking patterns in adult males. *J Consult Clin Psychol* 32:2–12, 1968.

52. Knight RG, Prout CT: A study of results in hospital treatment of drug addictions. *Am J Psychiatry* 108:303-308, 1951.

53. Boyd P, Layland WR, Crickmay JR: Treatment and follow-up of adolescents addicted to heroin. *Br Med J* 4:604–605, 1971.

54. Chein I: Psychological, social and epidemiological factors in drug addiction, in *Rehabilitating the Narcotic Addict.* US Vocational Rehabilitation Administration, Washington, DC, United States Government Printing Office, 1967, pp 53–66.

55. Wikler A, Rasor RW: Psychiatric aspects of drug addiction. *Am J Med* 14:566–570, 1953.

56. McGrath JH: Adolescent pill users. *Int J Addict* 5:173–182, 1970.

57. Ellinwood EH: Amphetamine psychosis: I. Description of the individuals and process. *J Nerv Ment Dis* 144:273–283, 1967.

58. Rosenberg CM: Young drug addicts: Background and personality. *J Nerv Ment Dis* 148:65–73, 1969b.

59. Tomsovic M: Hospitalized alcoholic patients: I. A two-year study of medical, social, and psychological characteristics. *Hosp Comm Psychiatry* 19:197–203, 1968.

60. Goldstein SG, Linden JD: Multivariate classification of alcoholics by means of the MMPI. *J Abnorm Psychol* 74:661–669, 1969.

61. Lawlis GF, Rubin SF: 16 PF study of personality in alcoholics. *Q J Stud Alcohol* 32:318–327, 1971.

62. Whitelock PR, Overall JE, Patrick JH: Personality patterns and alcohol abuse in a state hospital population. *J Abnorm Psychol* 78:9–16, 1971.

63. Skinner HA, Jackson DN, Hoffman H: Alcoholic personality types: Identification and correlates. *J Abnorm Psychol* 83:658–666, 1974.

64. McLachlan JFC: An MMPI discriminant function to distinguish alcoholics from narcotic addicts: Effects of age, sex, and psychopathology. *J Clin Psychol* 31:163–165, 1975.

65. Nerviano VJ: Common personality patterns among males: A multivariate study. *J Consult Clin Psychol* 44:104–110, 1976.

66. Skinner HA, Reed PL, Jackson DN: Toward the objective diagnosis of psychopathology: Generalizability of model personality profiles. *J Consult Clin Psychol* 44:111–117, 1976.

67. Hill HE, Haertzen CA, Glaser R: Personality characteristics of narcotic addicts as indicated by the MMPI. *J Gen Psychol* 62:127–139, 1960.

68. Glaser D, O'Leary V: *The control and treatment of narcotic use,* JD-5004-Parole Series. Justice Department, Washington, DC, United States Government Printing Office, 1966.

69. Berzins JI, Ross WF, Monroe JJ: A multivariate study of the personality characteristics of hospitalized narcotic addicts on the MMPI. *J Clin Psychol* 27:174–181, 1971.

70. Lasagna L, Mosteller F, von Felsinger JM, et al: A study of the placebo response. *Am J Med* 16:770–779, 1954.

71. Kornetsky C, Humphries O: Relationship between effects of a number of centrally acting drugs and personality. *Arch Neurol Psychiatry* 77:325–327, 1957.

72. Klerman GL, DiMascio A, Rinkel M, et al: *The influence of personality factors on the effects of phenotropic agents.* Paper presented at the meetings of Biological Psychiatry, San Francisco, Calif, 1958.

73. Heninger G, DiMascio A, Klerman GL: Personality factors in variability of response to phenothiazines. *Am J Psychiatry* 121:1091–1094, 1965.

74. Frostad AL, Forrest GL, Bakker CB: Influence of personality type of drug response. *Am J Psychiatry* 122:1153–1158, 1966.

75. Stanton JM: Group personality profile related to aspects of antisocial behavior. *J Criminal Law, Criminology, Police Science* 47:340–349, 1956.

76. Overall JE: MMPI personality patterns of alcoholics and narcotic addicts. *Q J Stud Alcohol* 34:104–111, 1973.

77. Lachar D, Gdowski CL, Keegan JF: MMPI profiles of men alcoholics, drug addicts and psychiatric patients. *J Stud Alcohol* 40:45–56, 1979.

78. Wexberg LE: Psychodynamics of patients with chronic alcoholism. *J Clin Psychopath* 10:147–157, 1949.

79. Hill HE: The social deviant and initial addiction to narcotics and alcohol. *Q J Stud Alcohol* 23:562–582, 1962.

80. Hill HE, Haertzen CA, Davis H: An MMPI factor analytic study of alcoholics, narcotic addicts and criminals. *Q J Stud Alcohol* 23:411–431, 1962.

81. Haertzen CA, Hill HE, Monroe JJ: MMPI scales for differentiating and predicting relapse in alcoholics, opiate addicts, and criminals. *Int J Addict* 3:91–106, 1968.

82. Rardin DR, Lawson TR, Kruzich DJ: Opiates, amphetamines, alcohol: A comparative study of American soliders. *Int J Addict* 9:891–898, 1974.

83. Black FW, Heald A: MMPI characteristics of alcohol- and illicit drug-abusers enrolled in a rehabilitation program. *J Clin Psychol* 31:572–575, 1975.

84. Aaronson BS: Age and sex influence on MMPI profile peak distributions in an abnormal population. *J Consult Psychol* 22:203–206, 1958.

85. Calden G, Hokanson JE: The influence of age on MMPI responses. *J Clin Psychol* 15:194–195, 1959.

86. McGinnis CA, Ryan CW: The influence of age on MMPI scores of chronic alcoholics. *J Clin Psychol* 21:271–272, 1965.

87. Gynther MD: White norms and black MMPI's: A prescription for discrimination? *Psychol Bull* 78:386–402, 1972.

88. Cavior N, Kurtzberg RL, Lipton DS: The development and validation of a heroin addiction scale with the MMPI. *Int J Addict* 2:129–137, 1967.

89. Hampton PJ: A psychometric study of drinkers. *J Consult Psychol* 15:501–504, 1951.

90. Holmes W: *The development of an empirical MMPI scale for alcoholics,* master's thesis. San Jose State University, 1953.

91. Hoyt DD, Sedlacek GM: Differentiating alcoholics from normals and abnormals with the MMPI. *J Clin Psychol* 14:69–74, 1958.

92. MacAndrew C: The differentiation of male alcoholic outpatients from nonalcoholic psychiatric outpatients by means of the MMPI. *Q J Stud Alcohol* 26:238–246, 1965.

93. Rich CC, Davis HG: Concurrent validity of MMPI alcoholism scales. *J Clin Psychol* 25:425–426, 1969.

94. Rosenberg N: MMPI alcoholism scales. *J Clin Psychol* 28:515–522, 1972.

95. Whisler RH, Cantor JM: The MacAndrew alcoholism scale: A cross-validation in a domiciliary setting. *J Clin Psychol* 22:311–312, 1966.

96. Rhodes RJ: The MacAndrew alcoholism scale: A replication. *J Clin Psychol* 25:189–191, 1969.

97. Rohan WP, Tatro RL, Rotman SR: MMPI changes in alcoholics during hospitalization. *Q J Stud Alcohol* 30:389–400, 1969.

98. Vega A: Cross-validation of four MMPI scales for alcoholism. *Q J Stud Alcohol* 32:791–797, 1971.

99. Hoffman H, Loper RG, Kammeier ML: Identifying future alcoholics with MMPI alcoholism scales. *Q J Stud Alcohol* 35:490–498, 1974.

100. Chang AF, Caldwell AB, Moss T: Stability of personality traits in alcoholics during and after treatment as measured by the MMPI: A one-year follow-up study. *Proceedings of the 81st Annual Convention of the American Psychological Association,* 1973, pp 387–388.

101. Jacobson GR: *Diagnosis and assessment of alcohol abuse and alcoholism,* DHEW PUBN (ADM) 75-228. US Department of Health, Education and Welfare, 1975.

102. Kranitz L: Alcoholics, heroin addicts and non-addicts: Comparisons on the MacAndrew Alcoholism Scale of the MMPI. *Q J Stud Alcohol* 33:807–809, 1972.

103. Surrey ES: *An investigation of the relationship of narcotic drug dependence, perceptual style, and stress.* Doctoral dissertation. George Washington University, 1967.

104. Witkin H: *Personality through perception: An experimental and clinical study.* New York, Harper, 1954.

105. Ballner WG: *The effects of emotional stress upon levels of anxiety and oral dependence as measured in the Holtzman Inkblot and TAT protocols of matched groups of heroin addicts, alcoholics, and controls,* doctoral dissertation. The Catholic University of America, 1975. (Xerox University Microfilms, Ann Arbor, Mich, 75-21, 532)

106. Rauchfleisch U: Psychodynamics of addictions (English translation). *Praxis Psychotherapie* 16:1–8, 1971.

107. Brien RL, Kleiman J, Eisenman R: Personality and drug use: Heroin, alcohol, methadrine, mixed drug dependency and the 16 PF. *Corrective Psychiatry and the Journal of Social Therapy* 18:22–23, 1972.

108. Zuckerman M, Neary RS, Brustman BA: Sensation-Seeking Scale correlates in experience (smoking, drugs, alcohol, "hallucinations," and sex) and preference for complexity [designs]). *Proceedings of the 78th annual convention of the American Psychological Association* 5:317–318, 1970. (Summary)

109. Zuckerman M, Bone RN, Neary R, et al: What is the sensation seeker? Personality trait and experience correlates of the Sensation-Seeking Scales. *J Consult Clin Psychol* 39:308–321, 1972.

110. Kaestner E, Rosen L, Appel P: Patterns of drug abuse: Relationships with ethnicity, sensation seeking, and anxiety. *J Consult Clin Psychol* 45:462–468, 1977.

111. Schwarz RM, Burkhart BR, Green SB: Turning on or turning off: Sensation seeking or tension reduction as motivational determinants of alcohol use. *J Consult Clin Psychol* 46:1144–1145, 1978.

112. Sutker PB, Archer RP, Allain AN: Drug abuse patterns, personality characteristics, and relationships with sex, race, and sensation seeking. *J Consult Clin Psychol* 46:1374–1378, 1978.

113. Carroll JFX, Santo Y, Klein MI: *A comparison of personality characteristics of male alcoholics and addicts as measured by the Personality Research Form,* unpublished manuscript, 1976. (Available from JFX Carroll, EHRC Psychology Dept, Eagleville, PA 19408.)

114. Nie MH, Hull CH, Jenkins JG, et al: *SPSS: Statistical Package for the Social Sciences,* ed 2. New York, McGraw-Hill, 1975.

115. Jackson DN: *Personality Research Form Manual.* Goshen, NY, Research Psychologist Press, 1967.

116. Fitts WH: *Tennessee Self Concept Scale Manual.* Nashville, Tenn, Counselor Recordings and Tests, 1965.

117. Stephens R, Levine S: The "street addict role": Implications for treatment. *Psychiatry* 34:351–357, 1971.

118. Cohen A, Barr HL, Ottenberg DJ: *Comparing heroin addicts and alcoholics on the Personality Research Form,* unpublished manuscript, 1978. (Available from HL Barr, EHRC Research Dept, Eagleville, PA 19408.)

5 Combined Alcohol-Drug Abuse and Human Behavior

Sidney Cohen

In a study of the effects of a single drug on behavior, the implications are manifold. Dosage levels, modes of administration, baseline states, the expectations of the subjects and of the investigators, the environments in which the drug is taken—all these variables, and others as well, make human psychochemical studies difficult and complex. When two or more drugs are used together or in sequence, the problems become magnified. Add to this analysis the vagaries of street drugs with their contaminants, adulterants, diluents, and haphazard quality and quantity control, the situation almost defies scientific scrutiny. Nevertheless, since polydrug use is notably prevalent and shows no signs of becoming less so,[1] an effort must be made to estimate the impact of multiple drug abuse.

Of all the polydrug patterns, those involving alcohol are the most frequently encountered and, perhaps, the most dangerous. The days when substance abusers were categorically labeled—as alcoholic, cokehead, hophead, pothead, and pillhead—seem to be rapidly disappearing. Instead, we are seeing people overinvolved with a primary substance of choice, but also using a variety of others, depending on availability, price, social situation, peer group usage, and the latest wisdom from the so-called underground press.

This chapter reviews the available literature on alcohol and other drug interactions in humans. It attempts to provide information on the psychophysiologic effects of specific combinations. Finally, the impact of multiple drug use on certain behaviors is described insofar as such effects are recorded in the literature.

A few definitions are provided here to ensure a uniform understanding of the material that follows:

Tolerance Tolerance is the need to increase the dose of certain regularly used drugs over time to achieve the same effects as originally desired. Narcotics, sedatives, minor tranquilizers, alcohol, and amphetamines produce tolerance.
Cross-tolerance After tolerance to a drug has developed, tolerance to others in the same class or in related classes will be present. A person who is tolerant to a barbiturate, for example, will be tolerant to other barbiturates, other sedatives, alcohol, and the minor tranquilizers.
Physical dependence (addiction) After tolerance has developed, the abrupt withdrawal of the drug will cause a pattern of symptoms called the withdrawal or abstinence syndrome. Tolerance, the withdrawal syndrome, and the desire or need to continue using the drug represent physical dependence.
Synergism When two drugs act similarly, they are synergistic.
Antagonism When two drugs have opposing effects, they are antagonists.
Additive When two drugs acting similarly are used together and the result is a simple summation of effect, they are considered to be additive.
Supra-additive (potentiation) When the effect of two synergistic drugs is greater than the sum of their doses, they are supra-additive.
Alcohol (ethanol) Alcohol is a general anesthetic. Like other anesthetics, this drug manifests an initial period of depression of the inhibitory control mechanisms that is experienced as behavioral stimulation.

Certain reasons for multiple drug ingestion exist. The most obvious one is to enhance the effects of the basic mind-altering substance used. Alcohol is a central nervous system (CNS) depressant. Using other classes of depressants (narcotics, sedatives, minor tranquilizers, or volatile solvents) along with alcohol, will, at a minimum, add to the depressed action. In certain instances the related drugs are supra-additive when used in combination with alcohol. These potentiating actions will be discussed further under the specific combinations of the depressant drugs with alcohol.

Another reason why more than one agent may be used is to counteract certain undesired effects of the basic psychochemical. Amphetamines make some users too tense and jittery ("wired up"), though euphoric. In such cases, alcohol is able to take the edge off the tension state.

At times, combinations of drugs are used when the preferred agent is not at hand, is of poor quality, or is too expensive. During a heroin "panic," codeine cough syrup, propoxyphene (Darvon), alcohol, and marijuana—alone or in combination—may be substituted.

Finally, there is multiple drug use for its own sake, without particular concern about the niceties of what the potpourri of chemicals will do. There is still an occasional person who will take anything and everything that is available. This mindless ingestion of a bewildering array of psychotropic drugs has been called "the garbage-head syndrome."

Combining psychopharmaceuticals can follow essentially two trends: 1) that of increasing central nervous system excitation or 2) that of increasing central nervous system depression. Alcohol, when combined with other depressants, can only increase sedation by an additive or potentiating action. When alcohol is combined with stimulants, the net effect might be to dampen or antagonize some of the undesired actions of the stimulant.

INTERACTIONS

The psychopharmacologic interactions of alcohol with other drugs are numerous. Three of the major levels of interplay that account for tolerance and cross-tolerance are:

1. The presence of the drug increases the amount of metabolizing enzymes responsible for their breakdown.
2. The response of the receptor cell, the neuron, for example, becomes more resilient to the action of the drugs by continued exposure.
3. The organism attempts to adapt to the presence of the mind-altering chemical by increasing its own self-surveillance and monitoring functions.

These adaptive efforts at subcellular, cellular, and behavioral levels result in a decreased effect of the drugs at their original dosage levels. If effects equivalent to the initial action of the drugs are desired, they can only be attained by consuming increased amounts of the agents.

Since the increased enzyme formation and the cellular and organismic adaptive efforts are often similar for other drugs in the same or related classes, cross-tolerance develops along with tolerance.

It should be emphasized that, for tolerance and cross-tolerance to occur, the drug or drugs must be consumed daily—usually a number of times a day for weeks or months.

Cross-tolerance is important in understanding the combined, *chronic* use of alcohol and related drugs. It means that a chronic alcoholic who is *not* actively drinking will be relatively resistant to an anesthetic agent, a sleeping potion, or a tranquilizer. On the other hand, when that person is actively drinking, the other depressant drugs will be additive or even supra-additive so that less than lethal amounts of each of the drugs can cause death by their total impact. This *acute* effect occurs because the amount of enzyme available to degrade all of the related drugs is insufficient to deal with large amounts of both, and their combined toxic effects can be lethal. This occurs even though larger-than-baseline amounts of the metabolizing enzyme have been induced by prior exposure to these drugs.

Cross-tolerance also has a therapeutic implication. It means that, in detoxifying an alcohol or barbiturate addict, any sedative, minor tranquilizer, or even alcohol could be used for the gradual elimination of these depressant drugs. Alcohol- and sedative-dependent people have learned that when their preferred drug happens to be in short supply, one of the other depressant drugs will avoid the depressant withdrawal syndrome (delirium tremens) that can develop when a person who has become tolerant to these substances suddenly stops their use. This procedure also applies to the use of methadone to maintain heroin-dependent persons. This is called cross-dependence—the ability of drugs of the same or related classes to suppress the abstinence syndrome. Since a narcotic like heroin is only distantly related to the sedative–minor tranquilizer–anesthetic group, heroin addicts prefer to search out codeine cough syrup, methadone, or some other narcotic when heroin becomes unavailable. These users will, nevertheless, take alcohol or sleeping pills to partially reduce the severity of their withdrawal symptoms.

Additional possible hazards of multiple drug use include:

- The addition or potentiation of depressant effects on the respiratory and cardiac regulatory centers in the brain may cause overdose and death.
- When all drugs are simultaneously discontinued, multiple withdrawal syndromes may emerge.
- Alcohol combined with intravenously-injected illicit drugs may create increased organ pathology, particularly of the liver.
- Increased perceptual and cognitive dysfunction may occur.

- The likelihood of behavioral problems is increased.
- As indicated by clinical experience, polydrug abuse tends to be more refractory to treatment than monodrug abuse.

The amount of brain dysfunction among polydrug-using individuals is a matter of some concern. In a collaborative study of five psychiatric centers, Grant et al[2] examined 151 polydrug users, 66 psychiatric patients, and 59 nonpatient subjects. The Halstead-Reitan Neuropsychological Test Battery, the MMPI, the WAIS, and an extensive drug use/medical history were obtained. Acute intoxication was ruled out by clinical observation and a urine drug screen. The results indicated that 37% of the polydrug group, 26% of the psychiatric patients, and 8% of the nonpatient controls were neuropsychologically impaired. On reexamination three months later, with diminished drug-alcohol usage, a quarter of the polydrug users had improved their Halstead-Reitan scores, indicating that a certain degree of reversibility of the mental deficits can occur in some instances.

Alcohol, the barbiturates, and certain other abused substances interfere with the therapeutic activity of many other classes of drugs. These interactions will not be considered here except to say that people on anticonvulsants, antibiotics, antidiabetic compounds, or anticoagulants should not drink or use barbiturates without their physician's approval. The very popular combination of alcohol and aspirin is capable of causing gastric bleeding because both are irritating to the stomach lining, and aspirin interferes with clotting. The use of large amounts of alcohol together with acetaminophen has been recently found to be hepatotoxic. There are also drugs such as disulfiram, metronidazole, and certain agents used for the treatment of diabetes that so interfere with the metabolism of alcohol that they induce an uncomfortable reaction—consisting of a severe flush, chest pain, a drop in blood pressure, and other symptoms. The alcohol-disulfiram reaction is used as deterrent therapy for certain patients with alcohol problems.

BIOBEHAVIORAL EFFECTS OF ALCOHOL-DRUG COMBINATIONS

Alcohol-Narcotic Combinations

Although ethanol and opiates do not potentiate each other, the combination of alcohol and heroin is a frequent cause of death[3]; they do have an additive effect. Moller[4] reported that a blood alcohol concentration (BAC) of 0.18% to 0.2% (definitely intoxicated) plus as little as 15 to 30 mg of morphine (an average dose) was fatal to nontolerant individuals. Similarly, patients on methadone maintenance who became secondarily addicted to alcohol have a mortality rate ten times that of maintenance patients not overinvolved with alcohol.[5] Baden[6] stated that combined narcotism and alcoholism was the cause of death for 10% of Manhattan heroin addicts who died between 1950 and 1961; more than 20% of these heroin addicts were found to have histories or autopsy findings of chronic alcoholism.

Another New York City autopsy report[7] stated that 30% of subjects assessed as opiate addicts were also alcoholics, and that 16% of the assessed alcoholics were also opiate addicts. The doubly-addicted individuals resembled those addicted only to alcohol in that they had more hospitalizations for illness during the year before their demise. Demographically, and as victims of homicide, they were similar to those addicted to narcotics only.

Doubly-addicted persons are particularly vulnerable to liver disease; they sustain the toxic effects of alcohol on the liver that progresses from fatty infiltration to alcoholic hepatitis and finally to cirrhosis. Overlaid upon this damage is the viral hepatitis that is introduced through nonsterile injections. The debilitating nutritional deficiencies common to the addicted add to the hepatic insufficiency. The reasons for the nutritional deficits can be summarized as follow: 1) poor intake of essential nutrients because of anorexia, vomiting, and the spending of available funds on alcoholic beverages; 2) poor assimilation of essential nutrients due to inflammatory changes in the gastrointestinal tract, and diarrhea; 3) increased utilization of certain vitamins in the metabolism of alcohol; and 4) increased loss of iron from hemorrhage.

Little has been written on the abuse of alcohol among patients on narcotic antagonists. Now that long-acting narcotic antagonists are being used on a wider scale, it might be predicted that those whose enjoyment of opiates has been blocked will turn to drink, and this may become a major problem when treatment involves naltrexone or other antagonists.

Experimental studies of human behavior using narcotics other than methadone and alcohol have not been found. This is understandable in view of ethical restrictions on administering the potentially dangerous combination of opiates plus alcohol to those who have never been addicted, or to exaddicts.

Propoxyphene is a commonly used analgesic with properties reminiscent of narcotics. In fact, its chemical structure resembles methadone. It has been known to be abused alone and in conjunction with alcohol. An average dose, 65 mg, was given to normal subjects along with a modest amount of alcohol (to effect a BAC of 0.05%) and matching placebos.[8] Each drug alone produced slight impairment on pursuit rotor performance, standing stability, and verbal tests. The combination resulted in an addictive effect that moderately impaired the subjects' test results.

Alcohol-Sedative Combinations

Combinations of alcohol and barbiturates are supra-additive because both chemicals compete for similar enzyme systems in order to complete their metabolic degradation. Milner[9] and Guptka and Kaford[10] reported that a BAC of 0.1% (the commonly accepted level of evidence of intoxication) and a blood barbiturate level of 0.5% (a third as high as the lethal level) had been fatal. Death occurs in such cases because barbiturate metabolism is inhibited, and the presence of the sedative in the organism is prolonged, producing coma and respiratory arrest at doses of each drug that ordinarily would not be fatal. This potentiation of alcohol and barbiturates holds only for the nontolerant person or for the chronic user of alcohol who is actively drinking. When an alcoholic stops drinking, the liver enzymes become available for barbiturate breakdown. In fact, because of long-term drinking, enzyme induction will have been stimulated, and barbiturates and certain other drugs are metabolized even more rapidly than usual; thus, a relative resistance to these drugs will occur. This phenomenon accounts for the difficulties in anesthetizing or sedating alcoholics who have recently stopped drinking when entering a hospital for surgery or detoxification from alcohol.

When the liver has been so extensively damaged from prolonged excessive drinking that it is unable to manufacture the metabolizing enzymes, alcohol and other abused substances are degraded inefficiently, and even small amounts recirculate and may cause severe intoxication.

Other depressant drugs that are potentiated by alcohol include paraldehyde, chloral hydrate, ether, and chloroform. One life-threatening combination is that of alcohol with carbon tetrachloride, a volatile solvent that was used until recently as a dry cleaner. Alcohol increases the solubility of carbon tetrachloride, which is highly toxic to the liver and kidneys.

Another substance requires special mention. Methaqualone in combination with alcohol has been described by Inaba et al[11] as occasionally producing stupor, coma, and respiratory depression. This reaction to the combination was known by its abusers as "luding out," or passing out. Although similar effects have been known with chloral-alcohol combinations (these are the reputed "knockout drops"), methaqualone-alcohol usage became increasingly popular until recently when methaqualone was placed in a more restrictive legal classification.

Sedatives and alcohol, sharing the same class, have many pharmacologic properties in common. They both produce tolerance and cross-tolerance; the withdrawal syndrome is identical (delirium tremens); they compete for similar enzyme systems in the liver; and they potentiate each others' effects. Many hypnosedative addicts abuse alcohol and vice versa, and the various patterns of abuse are similar: acute intoxication, binge usage, and continuous consumption of large amounts. It would be difficult to distinguish intoxication with barbiturates from that with alcohol except for the odor that accompanies the latter. Further, the use of both may be more common than is frequently realized. According to Devenyi and Wilson,[12] various studies indicate that 22% to 70% of certain alcoholic populations abuse barbiturates as shown by urine analyses.

Sedatives and alcohol are not infrequently taken together for suicidal purposes. DAWN IV data revealed that of 23,148 mentions of alcohol in combination, 32% were suicide attempts or gestures, 30% were used for psychic effects, 13% were consumed to maintain a state of dependence, and the remainder either gave no response or the motives were unknown (DAWN IV, 1976).

The combined effect of hypnosedatives and ethanol markedly worsens mental and motor performance. Loomis[13] demonstrated this in a driving simulator with secobarbital and alcohol. Phenobarbital can dramatically decrease reaction time in conjunction with alcohol intake.[14] Increased drowsiness and impaired motor function may persist for as long as 24 hours following the use of the combined drugs.[15]

The same deterioration of psychophysiological functioning occurs with chloral-ethanol combinations. Sellers et al[16] reported both substantial physiologic alterations and prolonged reaction time, impaired tracking, and vigilance decrements. Glutethimide and alcohol likewise have been shown to impair reaction times and other behavioral tasks more than either drug alone.[17]

Alcohol-Minor Tranquilizer Combinations

A large number of studies have proven that benzodiazepines such as chlordiazepoxide and diazepam do not potentiate alcohol.[18-21] Bernstein et al[22] could not even find an additive effect for diazepam and alcohol. There were no adverse drug interactions and no tolerance development. In fact, the benzodiazepines were considered preferred drugs for the acute treatment of the alcoholic patient. More recently, however, it has been found that large doses of the benzodiazepines have an effect on alcohol, and alcoholics given these agents over long periods during unsupervised treatment have taken them in increasing quantities.[23] An occasional fatality has been recorded when large amounts of these minor tranquilizers have been consumed with significant amounts of alcohol. Although the metabolism of the benzodiazepines is not disturbed by alcohol, the additive effect results from the CNS depressant activity of both drugs. Diazepam blood levels were higher when alcohol was consumed with the drug than when diazepam was used alone. A similar increase in blood levels was not found with a chlordiazepoxide-alcohol combination.[24]

The benzodiazepines are the most widely prescribed of all drugs; diazepam ranks first on national prescription audits. Bo et al[25] compared 74 motorists who had been hospitalized following accidents with a control group of 204 motorists who had not been in an accident. Studies revealed that 41.8% of those hospitalized retained alcohol in their blood, while only 1.5% of controls did so; and 9.5% of the injured drivers had diazepam in their blood, as compared with 2% in the control group. Of the accident group, a total of 10.8% had consumed both alcohol and diazepam before driving, while none of the control group had this combination on testing. This study demonstrated that the increased use of diazepam and its use concurrent with alcohol may significantly contribute to traffic accidents.

When low doses of a benzodiazepine are taken along with a small amount of alcohol, little or no impairment of functioning is detectable. In the higher dosage ranges behavioral deficits become obvious. Hughes et al[26] found no impairment on a pursuit tracking test or on a subjective symptom questionnaire after chlordiazepoxide and alcohol. Nor could Miller et al[19] or Bowes[18] detect any change on physiologic measures or on a digit symbol task. The administration of 5 mg of diazepam three times daily plus three ounces of 100-proof vodka did not impair subject performance on simple psychomotor tests.[27]

The majority of experimental studies do report decrements in various complex behavioral activities under combined benzodiazepine-alcohol use. Palva[28] demonstrated that chlordiazepoxide plus alcohol impaired short-term memory and learning. Performance deficits in eye control and standing steadiness were noted by Goldberg[29] from an alcohol-chlordiazepoxide combination. Burford et al[30] found that diazepam plus alcohol produced a more prolonged reaction time than did alcohol alone; their subjects were unaware of their impairment. In a driving test, Smiley et al[31] recorded a decreased ability of subjects on diazepam and alcohol to stop accurately. The subjects' wheel movements while under the influence of these substances were also different from normal wheel-handling patterns. Franks et al[32] and

Molander and Duvhok[33] used an extensive battery of sensory, perceptual, and psychomotor tasks, and detected increased deficits with benzodiazepine-alcohol combinations. Only the mood and subjective judgment of their test groups' condition were unchanged.

Five to 10 mg of diazepam plus either 0.5 or 0.8 g/kg of alcohol, plus appropriate placebos, were given to subjects who were tested with a complex reaction time test and tracking in a driving simulator. The subjects provided a rating of their performance.[24] Although the single drugs improved scores slightly, the combinations impaired performance. Moskowitz and Burns[34] gave their subjects 5 mg of diazepam plus enough alcohol to obtain a BAC of 0.07 mg%. In tests of tracking, visual search, information processing, and eye movement efficiency, the combination significantly impaired performance (these are psychomotor activities important in driving or operating machinery). Tested skills were performed less well under divided attention situations; this is also characteristic of driving behavior. On complex tasks like mirror tracking, time estimation, sorting, and letter cancellation, the drug combination resulted in definite impairment.[35] In assessing these studies, it should be remembered that the experimental benzodiazepine doses did not approach those ordinarily used by abusers of these drugs. Therefore, it would be expected that more serious deterioration of performance would occur among multiple drug abusers.

There seems to be little question that meprobamate-alcohol combinations worsen an array of psychomotor behaviors. Among other functions, oculomotor control, body steadiness, and fatigue[29]; reaction time and tracking[13]; simple arithmetic, visual illusion, and other psychophysiologic tests[36]; and time estimation, attention span, and alertness[37] all showed significant decrements. Meprobamate seems to have additive action when combined with alcohol as demonstrated with various performance tasks by Zirkle et al,[36] Reisby and Theilgaard,[37] and Goldberg.[38]

Alcohol-Marijuana Combinations

Marijuana and its active ingredient, delta-9-tetrahydrocannabinol (THC) have a sedative quality, and, when used with alcohol, an additional sedative effect is noted. Both alcohol and marijuana increase the pulse rate, and summative effects on heart rate have also been found. Whether cross-tolerance develops has not yet been clearly established.

The combination of cannabis and alcohol is a popular one, with the alcohol ordinarily being consumed in the form of sweet wines or beer.

It is believed that some of the combined use is the result of the relative lack of potency of the cannabis used in this country; the alcoholic beverage assumes the role of booster or enhancer of the effect of the marijuana.

There is general agreement that alcohol-marijuana combinations produce an additive decrement on various behavioral tasks. Chesher et al[39] indicated that standing steadiness, manual dexterity, and psychomotor skills were more impaired with the combined drugs than with either alone. Increased deficits in monitoring visual signals during a divided-attention task were observed by Macavoy and Marks.[40] Moskowitz[41] also demonstrated impaired vigilance, information processing, and oculomotor control. An additive effect on pursuit tracking patterns, mental arithmetic, and heart rate when cannabis and alcohol were combined was found by Manno et al.[42]

The mean tracking error score for a complex tracking task was higher for the combination of a low dose of THC (0.21 mg/kg) plus a low dose of alcohol (0.03% BAC) than for either a high dose of THC (0.88 mg/kg) or a higher dose of alcohol. The combination also caused greater pulse rate acceleration and conjunctival injection than either drug alone.[43]

Alcohol-Stimulant Combinations

Although amphetamines and ethanol are physiologic antagonists, only the depressant effects of alcohol are neutralized. When alcohol produces excitability, amphetamines will increase that excitability.[44] Seevers[45] found that alcohol-inebriated patients given amphetamines became overactive and more difficult to deal with.

Methylphenidate, an occasionally abused stimulant, is also a pharmacologic antagonist to ethanol; in fact, it has been used in the treatment of alcoholic coma.[46] However, behaviorally it can have a synergistic effect by increasing alcohol-induced hostility and paranoia.

In an early study Bruns[47] found that amphetamines counteracted the alcohol-induced impairment of psychomotor skills. Newman and Newman[48] could not completely confirm this. More recent investigations on pursuit tracking,[49] mental performance tasks,[50] and overall mental functioning[51] also reported a lack of improvement when amphetamines are added to alcohol; in fact, there was a further decrement. Wilson et al[52] used a complex battery of test situations and found that amphetamines did reverse the ethanol deficit on some mental performance tests but not on psychomotor skills, indicating that a very complicated interaction was taking place. It appears that the ability of a stimulant to counteract the decremental effects of alcohol is a function of task complexity. Simple tasks may be done better; difficult ones, worse.

Risk taking in the form of gambling, verbal productivity, and mood were examined under the influence of 45 g ethanol (approximately three drinks), of 15 mg dextroamphetamine, and of both substances combined.[53] Confidence, garrulousness, and mood elevation occurred with both drugs and also with the combination.

Kipperman and Fine[54] analyzed their group of amphetamine and alcohol abusers and found that they could be sorted into two types. Type A were older (28–45 years) and were long-term, primary alcoholics who also took amphetamines in order to remain awake and be able to drink more. Type B were younger (19–27 years) and were primary amphetamine abusers who used small amounts of alcohol to "level off" an amphetamine "run." After stopping their amphetamine spree, this group consumed large amounts of ethanol in order to fall asleep. Type A persons described a loss of inhibition, greater sociability, and increased self-esteem and euphoria on the combination. Type B individuals mentioned greater sensitivity, clearer thinking, and a lifting of depression. The alcohol made them more sociable. When sober, both groups were antisocial, were unable to hold jobs for long periods, had a record of truancy and disrupted family relationships, and had served time in jail. Both groups were moderately depressed and anxious, and seemed to be treating their depression with their abused drugs. A better treatment of their depression was seen as a potential therapeutic approach.

Since strong, black coffee is widely employed as a sobering-up device, it would be worth knowing whether caffeine actually does improve alertness and functioning of the intoxicated person. The conclusions of research are ambiguous. At times, improvement occurred; at others, there was a worsening of performance.[55-56]

Franks et al[32] found that people with a BAC of 0.09% were uninfluenced by 300 mg of caffeine (equivalent to about three cups of coffee). Standing steadiness with eyes closed, manual dexterity, numerical reasoning, perceptual speed, and verbal fluency were performed worse with the ethanol-caffeine combination. Reaction times were better after caffeine. No clear pattern of caffeine antagonizing the effects of alcohol could be observed; the BAC was unaffected by caffeine. The authors speculated that drinking drivers might feel more alert after caffeine and believe themselves to be recovered from their intoxication. Nevertheless, such drivers would remain handicapped in motor coordination and in search-and-recognition procedures.

Alcohol-Antidepressants

The tricyclic antidepressants, somewhat paradoxically, are not stimulants. Instead, some have a sedative quality, and for that reason are occasionally abused. They are also being used as suicide devices since they can induce major abnormalities of heart rhythm when swallowed in large amounts.

Amitriptyline given the evening before and prior to the testing in therapeutic doses, was combined with enough alcohol to produce a 0.08% BAC. Placebos were also used. The combination increased the error score during simulated driving, pursuit rotor, and dot tracking. Even in ordinary doses, the interaction can be hazardous.[57] Seppola et al[58] administered either amitriptyline or doxepin in combination with alcohol in 0.5 g/kg quantities. The combination produced slight impairment of choice reaction time, coordination tests, attention, and tracking tasks. The subjects' assessment confirmed that they were impaired. These changes occurred even though the antidepressants were given for ten days prior to the test day; the expectation would be that tolerance to any sedative effect would have occurred.

Alcohol-Antihistamine Combinations

Many antihistamines such as diphenhydramine, chlorpheniramine, and others produce drowsiness in many people. For that reason these substances appear in over-the-counter sleeping preparations, which are known to be occasionally abused. When combined with alcoholic beverages, the sedative effects of both are additive. Linnoila[59] detected a significant worsening of coordination when diphenhydramine–alcohol mixtures were ingested. Pursuit tracking was impaired with a similar combination.[26]

Alcohol-Nicotine Combinations

In view of recent work suggesting that alcoholics who smoke have cancers of the head, neck, and esophagus more frequently than nonsmoking alcoholics or nonalcoholic smokers, the cigarette-alcohol relationship has become a public health issue.[60] The basis for the increased incidence of these malignancies is ethyl alcohol's ability to increase the solubility and absorption of coal tars from tobacco. Elber[61] and Lickint[62] have evidence that the combination worsens dexterity and mentation.

Heavy drinkers also tend to be heavy smokers,[63] although just why these patterns should coexist is not clear. Nicotine does not reverse the undesired effects of excessive alcohol consumption and certainly not alcohol-impaired performance.[64] It may be that the desirable effect occurs on a subjective level, or that both heavy smoking and drinking are strongly conditioned behaviors.

PATTERNS OF MULTIPLE DRUG ABUSE

The NDACP Study

The National Drug/Alcohol Collaborative Project (1977) obtained significant data pertaining to concurrent multiple drug abuse. It should be noted that the NDACP sample of patients in treatment was neither random nor representative, and that information was derived from self-reports. Table 5-1 shows the numbers of regular users of 14 different substances which altered (boosted, balanced, counteracted, or sustained) the effects of their primary drug. Table 5-2 lists the major substance used to alter the effects of the primary drug. Table 5-3 shows the number of regular users of a drug who reported that they substituted another drug when the primary drug was not available. Table 5-4 notes the drug that was most commonly employed as a substitute when the primary drug could not be obtained.

From these and other patient-derived data, the following results were obtained. It should be pointed out that the analyses of substance-abuse patterns were conducted for two time frames: the three months previous to admission and the entire drug-using career previous to admission.

1. NDACP subjects were primarily multiple-substance abusers.
2. Alcohol was the substance abused most often by both single- and multiple-drug abusers.
3. The greatest number of subjects classified as career multiple-substance abusers had used alcohol and only one other substance (per the NDACP Final Report). In descending order of frequency, the other drugs were marijuana/hashish (n = 159); minor tranquilizers (n = 75); heroin (n = 19); amphetamines (n = 14); and barbituates (n = 10).

Table 5-1
The Number and Percentage of Regular Drug Users of Each Drug Category Who Reported Altering the Effects of That Drug

	Alcohol	Heroin	Other opiates	Amphetamines	Barbiturates	Minor Tranquilizers	Marijuana/Hashish	Illicit methadone	Cocaine	Hallucinogens	Inhalants	OTC	Anti-depressants	Major tranquilizers
Altered	497	320	130	234	161	149	501	97	223	205	56	37	36	21
	53%	71%	66%	72%	75%	55%	72%	72%	77%	72%	36%	54%	80%	57%
Did not alter	434	132	66	92	51	124	154	37	67	80	101	32	9	16
	47%	29%	34%	28%	24%	45%	24%	28%	23%	28%	64%	46%	20%	43%

Source: NDACP Final Report, 1977.

Table 5-2
The Single Substance Most Commonly Used to Alter Each Other Substance

Initial Substance	Substance Used to Alter Effect of Initial Substance
Alcohol	Marijuana
Heroin	Marijuana
Other opiates	Alcohol
Amphetamines	Alcohol
Barbiturates	Alcohol
Minor tranquilizers	Alcohol
Marijuana/hashish	Alcohol
Illegal methadone	Heroin
Cocaine	Heroin
Hallucinogens	Marijuana

Source: NDACP Final Report, 1977.

4. Most multiple-substance abusers partook of alcohol and one other substance during the three months before admission. In descending order of abuse the other drugs were marijuana ($n = 209$); minor tranquilizers ($n = 61$); and heroin ($n = 41$). (NDACP Final Report.)
5. For multiple-drug abusers who had abused alcohol and two other drugs, the pattern usually consisted of alcohol, marijuana, and another drug; alcohol, a minor tranquilizer, and another drug; and alcohol, heroin, and another drug.
6. When alcohol and three other drugs were abused, the combinations included depressants and stimulants, and all depressants. A small group used stimulants, depressants, and hallucinogens.
7. The majority of subjects in every substance-abuse category (except inhalants) reported using one or more drugs to alter the substances already taken; more than 75% of the regular users of barbiturates, marijuana, cocaine, and antidepressants reported such use. The two drugs most commonly used for this purpose were alcohol and marijuana.
8. The two drugs that were most frequently substituted for were heroin and illegal methadone. Alcohol and inhalants least frequently required substitute drugs.

Table 5-3
Number and Percentage of Regular Users of a Drug Who Reported Substituting Other Drugs for It

	Alcohol	Heroin	Other Opiates	Amphetamines	Barbiturates	Minor tranquilizers	Marijuana/Hashish	Illicit Methadone	Cocaine	Hallucinogens	Inhalants
Substituted	299	269	92	119	115	112	294	92	138	112	54
	31%	58%	48%	37%	48%	32%	44%	69%	48%	39%	29%
Did not substitute	675	189	101	205	123	163	372	41	151	174	109
	69%	42%	52%	63%	52%	68%	56%	31%	52%	61%	71%

Source: NDACP Final Report, 1977.

Table 5-4
Most Commonly Reported Substitute for Each Substance

Initial Substance	Substitute
Alcohol	Marijuana
Heroin	Other opiates
Other opiates	Heroin
Amphetamines	Marijuana
Barbiturates	Minor tranquilizers
Minor tranquilizers	Alcohol
Marijuana/hashish	Alcohol
Illegal methadone	Heroin
Cocaine	Heroin
Hallucinogens	Marijuana

Source: NDACP Final Report, 1977.

Alcohol and marijuana were the drugs most frequently used as substitutes.

9. The amount of alcohol used by all subjects was high, averaging 11.4 ounces of whiskey a day.
10. The greatest amount of alcohol consumed was reported by those who had abused seven or more substances at some time during their substance-abuse careers.
11. The career alcohol–other opiate abusers had the highest alcohol consumption, followed by alcohol–minor tranquilizers, alcohol-marijuana, and alcohol-heroin.
12. During the past three months the descending order for alcohol–other drug intake was: alcohol-hallucinogens, alcohol-inhalants, alcohol-barbiturates, alcohol-amphetamines, and alcohol-cocaine.
13. Those who abused alcohol and marijuana concurrently reported smoking more marijuana than those using marijuana alone.
14. The greatest number of adverse effects from alcohol were reported by the users of alcohol-amphetamines, followed by the alcohol–minor transquilizer and the alcohol-barbiturate groups.

The Tuckfeld Study

In an effort to determine drug usage trends in clients attending alcohol treatment facilities, Tuckfeld et al[64] interviewed the service deliverers in six representative clinics. A widespread use of drugs, primarily opiates, was reported, and the service deliverers mentioned the following trends.

1. Persons under 25 were increasingly using alcohol either in combination with other drugs or when illicit drugs were inaccessible. Marijuana use was reported as a norm for this age group.
2. For persons under 30 who used drugs, males outnumbered females three to one. Besides alcohol, the drugs most often reported were marijuana and other psychotropic drugs. Barbiturates and amphetamines were also commonly used, although barbiturate use was reported to be on the decline. Hallucinogens were used more frequently by this age group than by the over-30 group.
3. Females over 30 years of age primarily abuse minor tranquilizers and other psychotropics. This was particularly pronounced for subjects from middle- and upper-middle-class households. Males over 30 were reported as primarily using psychotropics, and at one data collection site amphetamines were extensively used.
4. The particular substance used was more likely to be a function of drug accessibility and subcultural norms than of socioeconomic status or racial/ethnic group. Persons of lower socioeconomic status, however, rarely tended to use drugs other than alcohol unless they had recently visited a free clinic or public hospital. The drugs primarily used were psychotropics.
5. Consumption patterns were either conjoint or sequential. Conjoint use characterized persons who consumed drugs other than alcohol for recreational purposes. Sequential use was generally associated with attempts to improve daily social functioning or to cope with sobriety until a return to alcohol. Over-the-counter drugs were used primarily for self-medication.
6. It was estimated that 30% to 60% of all clients at alcoholism treatment facilities were using other substances at intake. Of these, half were abusing such drugs (abuse as defined as the nonmedical use of prescription drugs or the use of illicit drugs).
7. Persons under 30 were believed to have the highest incidence of multiple-drug use. Females had a higher rate of multiple-drug use than males.

8. Public inebriates were reported to have a lower rate of illicit multiple-drug use than did populations from higher socioeconomic levels.

PSYCHOSOCIAL EFFECTS OF ALCOHOL-DRUG COMBINATIONS

Drug combination effects on human behavioral patterns have been explored only recently; however, certain investigations have special importance for this chapter. These will be discussed according to specific behaviors and their modification by alcohol combined with other substances.

Alcohol-Drug Effects on Sexuality and Violence

Shakespeare's comment in *Macbeth* citing the negative impact that drinking has on the sexual response remains valid; in fact, it is being confirmed by recent investigations. Gebhard[65] suggests that small amounts of alcohol or related drugs may initiate sexual activity by lessening inhibitions and producing euphoria. The recent finding that chronic drinkers have an elevated level of luteinizing hormone may also help explain the increased sexual arousal. However, larger amounts of the same depressant substances tend to decrease sexual ability. Recent work has demonstrated that chronic heavy drinkers have a decreased plasma testosterone level, which may contribute to impotence. There are few experimental studies on the impact of alcohol on sexual arousal, and no investigations of alcohol conjointly administered with another drug in the area of sexuality.

The well-known propensity of alcohol for unleashing aggressive verbal and physical behavior is paralleled by the sedatives because of their similar disinhibiting properties. In a commentary on drugs and crime, Tinklenberg[66] notes that intradrug interactions have exceedingly complex behavioral effects, and that combinations have not been investigated in regard to their ability to induce violence. It may well be that the social and cultural contexts in which psychochemicals are used tend to determine their aggressive component.

Alcohol's direct relation to violence and violent crime is considered far greater than that of any other drug. Fitzpatrick[67] has found that alcohol-barbiturate combinations also produce high levels of assaultive behavior.

Information obtained in the NDACP[68] study provides further data on the role of both alcohol and drugs in violent acting out. In answer to the question, "Have you ever gotten angry or violent and seriously injured someone while under the influence of drugs, alcohol or both?", 39% of the regular alcohol users indicated that they had done so. Comparable confirming percentages for heroin users were 42%; for amphetamine users, 45%; users of barbiturates, 55%; and users of marijuana, 46%. Of those respondents who had been in automobile accidents, the following percentages had used drugs and/or alcohol immediately prior to the experience: heroin users, 36%; amphetamine users, 43%; users of barbiturates, 35%; users of marijuana, 36%; and alcohol users, 38%.

Alcohol Use and Drug-Taking Behavior

The DARP (Drug Abuse Reporting Program) data for patients who had been in federal drug treatment programs showed that 23% drink the equivalent of more than eight ounces of 80-proof liquor daily.[69] Day-to-day variations on this average were related to negative and positive life experiences. Fifteen percent drank less and 22% drank more when illicit drugs were used. Among the heavier drinkers the mean use of opioids, nonopioids, and marijuana decreased while drinking, with less opioid use being statistically significant ($p = .001$). When less alcohol was consumed by these former patients, the means for all illicit drug use increased; however, only opioid use increased significantly ($p = .03$). For these people alcohol seemed to be acting as a substitute for illicit drug use, particularly opioids. Life-stress situations and the persuasion of friends accounted for some of the variance. Alcohol consumption was less for patients in treatment than for those no longer in treatment. It was pointed out that excessive alcohol use sometimes predated the drug abuse. Drinking tended to decrease when opiate use started. Those who were heavy drinkers before their opiate dependence tended to continue that behavior while on methadone maintenance.

Suicide and Alcohol-Drug Combinations

The DAWN IV data (May 1975–April 1976) have already been mentioned.[70] This report indicated that suicidal attempts or gestures were the leading reason that patients who had taken alcohol in combination with other drugs appeared at crisis centers, emergency rooms, or the morgue. Of 23,148 mentions of those using alcohol in combination, 32% were suicide attempts/gestures, 30% had consumed the substances for the psychic effects involved, and 13% had used the combination to maintain their dependence. The remainder gave no response.

For purposes of suicide the intake of alcohol with other depressants is pharmacologically rational since the combination produces additive or potentiating effects. Then, too, depressive tendencies in substance users may be related to such gestures: chronic alcoholics are more prone to attempt suicide than nonalcoholics.[71] Further, about one-quarter of suicides are committed by alcoholics.

The NDACP data[68] address these relationships. Two-thirds of those who initiated suicide attempts were under the influence of a psychotropic drug at the time. More than half used a psychochemical or drug combination in the attempt to end their lives. About one-quarter used barbiturates, some 15% used alcohol, and the agents used by the rest were distributed among many other drugs or drug combinations.

Alcohol-Depressant Combinations and Sleep

Acute alcohol intoxication is known to distort normal sleep patterns. Rapid-eye-movement (REM) sleep is diminished, particularly during the first half of the night, while slow-wave sleep is increased (sleeping pills have a similar effect on the sleep EEG). Chronic alcoholism can also be associated with extreme sleep disturbances with a considerable REM deficit occurring. When the amount consumed decreases or is eliminated, a REM rebound with accompanying nightmares develop, resulting in fragmented sleep patterns and insomnia. At times, transitional states occur in which individuals cannot identify whether they are having nightmares or are awake and hallucinating. During withdrawal REM sleep may increase, consuming up to 100% of sleep time.[72] The hypnosedatives, the minor tranquilizers, and related drug groups do not improve this sleep disturbance of the alcoholic; in fact, combined alcohol-sedative dependencies produce similar sleep disturbances.

Alcohol is often used by alcoholics and nonalcoholics to procure sleep. A dramatic impairment of sleep, on the other hand, occurs early in the withdrawal phase when people verging on delirium tremens are afraid to close their eyes because they start hallucinating.

Gross and Hasty[72] describe the sleeplessness that accompanies the alcohol withdrawal state as consisting, in part, of an intense fear of not sleeping that approaches insomnophobia. Alcoholics so affected may try to drink themselves to sleep, to swallow whatever sleeping medication is at hand, or both.

Subjective Effects Reported by Alcohol and Drug Users

The NDACP questionnaire[68] inquired into the subjective effects of alcohol and specific drugs in those people who were users of alcohol and/or drugs. The question covered 37 areas of emotional and behavioral changes under the influence of either alcohol or the other substance. The most common combinations in the sample were: alcohol and heroin; alcohol and other opiates; alcohol and barbiturates; alcohol and minor tranquilizers; alcohol and marijuana; and alcohol and amphetamines.

When compared to all other categories, alcohol was named as a confusion-causing agent more frequently than any other substance. According to the respondents, alcohol also produced much more anger and fewer feelings of peace, caused a greater loss of control, and was less effective in awareness in comparison to other agents.

CONCLUSION

In reviewing the empirical evidence, it can be concluded that joint substance abuse is of considerable significance. Such usage may relate to ignorance of the pharmacologic facts, as with the combination of amphetamines or antidepressants with social drinking. Other cases may involve the addict who consciously combines intoxicating levels of ethanol and large amounts of psychoactive drugs. While the social "misuser" of substances may suffer impairment of abilities related to normal functioning and/or driving, chronic abusers may pose considerable risks to themselves and to others. Though experimental studies concerning the latter subject are understandably lacking, clinical and autopsy reports serve to underscore the hazardous nature of unmoderated alcohol/drug use.

The biobehavioral evidence surveyed in this chapter indicates that almost all classes of drugs have additive or supra-additive effects when combined with alcohol. Even the stimulants can increase the excitation phase of drinking and may intensify paranoid ideation. Studies attempting to show that stimulants reverse the psychomotor decrements of the alcohol-intoxicated person have not been conclusive. Therefore, efforts to treat intoxication with stimulants may result in unanticipated consequences.

Another point that should be reemphasized is the biphasic nature of certain drug-alcohol interactions. Acute alcohol intake along with sedative consumption can lead to the potentiation of depressant effects. On the other hand, the chronic alcoholic who stops drinking is more resistant to sedatives, and this is common hospital experience in attempting to quiet patients during alcohol withdrawal.

The impact of most drugs when combined with alcohol is to increase dysfunctional and antisocial behavior. Research indicates that impaired sexual functioning, disinhibition of aggressive behavior, suicide attempts/gestures, and sleep disturbances result from conjoint substance abuse. These findings are supplemented by the subjective testimony of addicts in treatment who report increased anger and loss of control/self-awareness under the influence of alcohol. Data bases such as the DARP, which suggests that increased alcohol intake is often associated with decreased drug intake following treatment, are of special interest to clinic personnel concerned with these problem behaviors.

The National Institute on Alcohol Abuse and Alcoholism has prepared a prototype prescription form that can be given to those having difficulty abstaining from drinking; it mentions that many unwished-for interactions can occur when certain therapeutic drugs are mixed with alcoholic drinks. In view of the bio- and sociobehavioral problems described in this chapter, it would seem that pharmacists, physicians, and clinic personnel should be prepared to similarly advise their clients/patients on the possible ill effects of combined substance use. More importantly, they should stress that the combination of two or more mind-altering agents produce not only successively greater impairments, but also tend to increase their unpredictability. Future research should address the present gaps in current knowledge—and serve to focus public attention on the problem as well.

REFERENCES

1. O'Donnell JA, Voss HL, Clayton RR, et al: *Young Men and Drugs—A Nationwide Survey.* NIDA Research Monograph 5, Rockville, Md, National Institute on Drug Abuse, 1976.

2. Grant I, Mohns L, Miller M, et al: The neuropsychological effects of polydrug use: Results of the national collaborative study. *Arch Gen Psychiatry* 33:973–978, 1976.

3. Erola R, Alha A: Synergism of alcohol and sedatives. *Deutsch Zeitsch Ges Gericht Med* 53:201–210, 1963.

4. Moller KD: Death from therapeutic doses of morphine or morphine-scopolamine in persons affected by alcohol or barbituric acid. *Bull Narcotics* 5:11–19, 1953.

5. Roizen L: *Interactions of Methadone and Ethanol.* Presented at a meeting of the Eastern Section of the American Psychiatric Association, New York, 1969.

6. Baden MM: Homicide, suicide and accidental death in Manhattan. *Hum Pathol* 3:91–95, 1972.

7. Haberman PW, Bader MM: Drinking, drugs and death. *Int J Addict* 9:761–773, 1974.

8. Kiplinger GF, Sokol G, Rhode BE: Effects of combined alcohol and propoxyphene on human performance. *Arch Int Pharmacodyn* 213:175–180, 1974.

9. Milner G: Interaction of barbiturates, alcohol and some psychotropic drugs. *Med J Aust* 1:1204–1207, 1970.

10. Guptka RC, Kaford J: Toxicological statistics for barbiturates, other sedatives, and tranquilizers in Ontario: A ten year survey. *Can Med Assoc J* 94:863–865, 1966.

11. Inaba DS, et al: Methaqualone abuse: Luding out. *JAMA* 224:1505–1509, 1973.

12. Devenyi P, Wilson M: Barbiturate abuse and addiction and their relationship to alcohol and alcoholism. *Can Med Assoc J* 215:215–218, 1971.

13. Loomis TA: Effects of alcohol on persons using tranquilizers, in Harvard JDD (ed): *Alcohol and Road Traffic.* Proceedings of the Third International Conference on Alcohol and Road Traffic, London, British Medical Association, 1963, pp 119–122.

14. Joyce CRB, Edgecombe PCE, Kennard DA, et al: Potentiation by phenobarbitone of effects of ethyl alcohol on human behavior. *Br J Psychiatry* 105:51–60, 1969.

15. Doenicke A, Kugler J: Investigations after barbiturate medication and additional ingestion of alcohol in the course of 24 hours: Establishing the barbiturate level in the serum: Tests of liver function and psychodiagnostic tests. *Aktuel Probl Chir* 2:134–148, 1965.

16. Sellers EM, Carr G, Bernstein JG: Interaction of chloral hydrate and ethanol in man. *Clin Pharmacol Ther* 134:50–58, 1972.

17. Mould GP, Curry SH, Binns TB: Interactions of glutethimide and phenobarbitone with ethanol in man. *J Pharm Pharmacol* 24:894–899, 1972.

18. Bowes HA: The role of Librium in an outpatient psychiatric setting. *Dis Ner Sys* 21:20–22, 1960.

19. Miller AJ, D'Agostino A, Minsky R: Effects of combined chlordiazepoxide and alcohol in man. *Q J Stud Alcohol* 24:9–13, 1963.

20. Votarova Z, Dyntarova H: Comparison of the chlormethiazole and diazepam effects on alcohol induced changes of EEG and behavior in rats, in *Collegium Internationale Neuropsychopharmacologium.* (CINP) Abstracts II. International Conference on Neuropsychopharmacology, Prague, 1970.

21. Vaapatalo H, Karppanen H: Combined toxicity of ethanol with chlorpromazine, diazepam, chlormethiazole of pentobarbital in mice. *Agents Actions* 1:43–45, 1969.

22. Bernstein ME, Hughes FW, Forney RB: The influence of a new chlordiazepoxide analogue on human and mental and motor performance. *J Clin Pharmacol* 7:330–335, 1967.

23. Hollister LE: Valium: A discussion of current issues. *Psychosomatics* 18:44–58, 1977.

24. Linnoila M, Mattila MJ: Drug interaction on psychomotor skills related to driving. *Eur J Clin Pharmacol* 5:186–194, 1973.

25. Bo O, Haffner JFW, Langard O, et al: Ethanol and diazepam as causative agents in road traffic accidents, in Israelstam S, Lambert S (eds): *Alcohol, Drugs and Traffic Safety.* Toronto, Addiction Research Foundation, 1975, pp 439–448.

26. Hughes FW, Forney RB, Richards AB: Comparative effects in human subjects of chlordiazepoxide, diazepam and placebo on mental and physical performance using a pursuit meter with and without low levels of alcohol. *Clin Pharmacol Ther* 5:139–145, 1964.

27. Lawton MP, Cann B: The effects of diazepam (Valium) and alcohol on psychomotor performance. *J Nerv Ment Dis* 136:550–554, 1963.

28. Palva ES: Effect of active metabolites of chlordiazepoxide and diazepam alone and in combination with alcohol on psychomotor skills related to driving. *Modern Problems of Pharmacopsychiatry* 2:79–84, 1976.

29. Goldberg L: Effects and aftereffects of alcohol, tranquilizers and fatigue on ocular phenomena, in Harvard JDD (ed): *Alcohol and Road Traffic,* Proceedings of the Third International Conference on Alcohol and Road Traffic, London, British Medical Association, 1963, pp 123–135.

30. Burford E, French IW, LeBlack AE: The combined effects of alcohol and common psychoactive drugs: I. Studies of human pursuit tracking capability, in Israelstam S, Lambert S (eds): *Alcohol, Drugs and Traffic Safety.* Toronto, Addiction Research Foundation, 1975, pp 423–431.

31. Smiley A, et al: The combined effects of alcohol and common psychoactive drugs: II. Field studies with an instrumented automobile, in Israelstam S, Lambert S (eds): *Alcohol, Drugs, and Traffic Safety.* Toronto, Addiction Research Foundation, 1975.

32. Franks HM, Starmer GA, Chesher GB, et al: The interaction of alcohol and delta-9-tetrahydrocannabinol in man: Effects of psychomotor skills related to driving, in Israelstam S, Lambert S (eds): *Alcohol, Drugs and Traffic Safety.* Toronto, Addiction Research Foundation, 1975, pp 461–466.

33. Molander L, Duvhok L: Acute effects of oxazepam and methylperone, alone and in combination with alcohol on sedation: Coordination and mood. *Acta Pharmacol Toxicol* 38:145–160, 1976.

34. Moskowitz H, Burns M: The effects of alcohol and Valium singly and in combination upon driving related skills performance. Personal communication, 1977.

35. Morland J, Setekleiv J, Haffner JFW, et al: Combined effects of diazepam and ethanol on mental and psychomotor functions. *Acta Pharmacol Toxicolog* 34:5–15, 1974.

36. Zirkle GA, McAtee OB, King PD, et al: Meprobamate and small amounts of alcohol: Effects on human ability, coordination and judgment. *JAMA* 173:1823–12825, 1960.

37. Reisby N, Theilgaard A: The interaction of alcohol and meprobamate in man. *Acta Psychiatr Scand* (suppl)5:204–208, 1968.

38. Goldberg L: Effects of ethanol in the central nervous system, in Popham RE (ed): *Alcohol and Alcoholism.* Toronto, University of Toronto Press, 1970, pp 42–56.

39. Chesher GB, Franks HM, Hensley VR, et al: The interaction of ethanol and delta-9-tetrahydrocannabinol in man: Effects of perceptual, cognitive and motor functions. *Med J Aust* 2:159–163, 1976.

40. Macavoy MG, Marks DF: Divided attention performance of cannabis users and non-users following cannabis and alcohol. *Psychopharmacologia* 44:147-152, 1975.

41. Moskowitz H: Marihuana and driving. *Accident Anal Prev* 9:21-26, 1976.

42. Manno JE, Kiplinger GF, Scholz M, et al: The influence of alcohol and marihuana on motor and mental performance. *Clin Pharmacol Ther* 12:202-211, 1971.

43. Hansteen RW, Miller RD, Lonero L et al: Effects of cannabis and alcohol on automobile driving and psychomotor tracking. *Ann NY Acad Sci* 282:240-256, 1976.

44. Weiss B, Laties VG: Effects of amphetamine, chlorpromazine, pentobarbital and ethanol on operant response conditioning. *J Pharmacol Exp Ther* 144:17-23, 1964.

45. Seevers MJ: Amphetamines and alcohol. *JAMA* 184:843, 1963.

46. Horvath D: To what extent and in what form do some so-called "fashionable" drugs affect the action of alcohol. *Orv Hetil* 104:2233-2236, 1963.

47. Bruns OI: Pervitin: Pharmacological and clinical aspects. II. Pervitin: The question of performance enhancement. *Fortsch Ther* 17:37-44, 90-100, 1941.

48. Newman HW, Newman EJ: Failure of dexedrine and caffeine as practical antagonists of the depressant effects of ethyl alcohol in man. *Q J Stud Alcohol* 17:406-410, 1956.

49. Brown DJ et al: Effects of dextroamphetamine and alcohol on attentive motor performance in human subjects, in Harger RH (ed): *Alcohol and Traffic Safety.* Bloomington, Ind, Indiana University Press, 1966, pp 215-219.

50. Kaplan HL, Forney RB, Richards AB, et al: Dextro-amphetamine, alcohol, and dextro-amphetamine-alcohol combination and mental performance, in Harger RN (ed): *Alcohol and Traffic Safety:* Proceedings of the Fourth International Conference on Alcohol and Traffic Safety at Indiana University, December 6-10, 1965. Bloomington, Ind, Indiana University Press, 1966, pp 211-214.

51. Hughes FW, Forney RB: Comparative effect of three antihistamines and ethanol on mental and motor performance. *Clin Pharmacol Ther* 5:414-421, 1964.

52. Wilson L, Taylor JD, Nash CW, et al: The combined effects of ethanol and amphetamine sulfate on performance in human subjects. *Can Med Assoc J* 94:478-484, 1966.

53. Hurst PM, Radlow R, Chubb NC, et al: Effects of alcohol and D-amphetamine upon mood and volition. *Psychol Rep* 24:975-987, 1969.

54. Kipperman A, Fine EW: The combined abuse of alcohol and amphetamines. *Am J Psychiatry* 131:1277-1280, 1974.

55. Alsatt RL, Forney RB: Performance studies in rabbits, rats and mice after administration of L-methyl-xanthene, singly and with ethanol. *Fed Proc* 30:568, 1971.

56. Nash H: Psychological effects and alcohol-antagonizing properties of caffeine. *Q J Stud Alcohol* 27:727-734, 1966.

57. Landauer AA, Milner G, Patman J: *Science* 163:1467-1468, 1969.

58. Seppola TM, et al: Effect of tricyclic antidepressants and alcohol on psychomotor skills related to driving. *Clin Pharmacol Ther* 17:515-522, 1975.

59. Linnoila M: Effects of antihistamines, chlormethazone and alcohol on psychomotor skills related to driving. *Eur J Clin Pharmacol* 5:247–254, 1973.

60. Alcohol and health. *Second Special Report to the U.S. Congress* No. 77-9099, Washington, DC, US Government Printing Office, 1976.

61. Elber H: Demonstration of effects of caffeine on blood alcohol level and drunkenness. *Beitr Gerichtl Med* 15:14–25, 1939.

62. Lickint F: The evoking of abnormal alcohol reactions by drugs. *Therapiewoche* 7:414–418, 1957.

63. Simon W, Lucero RJ: Consumption of metholated cigarettes by alcoholics. *Dis Nerv Sys* 21:213–214, 1960.

64. Tuckfeld BS, McLeary KR, Waterhouse DJ: *Multiple Drug Use Among Persons with Alcohol-Related Problems.* Springfield, Va, National Technical Information Service, 1975.

65. Gebhard PJ: Situational factors in affecting human sexual behavior, in Beach FA (ed): *Sex and Behavior.* New York, John Wiley & Sons, 1965.

66. Tinklenberg J: Drugs and crime, in *Drug Use in America: Problem in Perspective,* Appendix, vol 1, in technical papers of the second report of the National Commission on Marihuana and Drug Abuse. Washington, DC, US Government Printing Office, 1973.

67. Fitzpatrick JP: Drugs, alcohol and violent crime. *Addict Dis* 1:353–367, 1974.

68. Carroll JFX (ed): *National Drug/Alcohol Collaborative Project (NDACP): Final Report, 1977.* Eagleville, Pa, Eagleville Hospital and Rehabilitation Center, 1977.

69. Simpson DD, Lloyd MR: *Alcohol and Illicit Drug Use.* Institute of Behavioral Research report 77-2. Fort Worth, Tex, TCU, 1977.

70. *DAWN IV.* Rockville, Md, NIDA, 1976.

71. Goodwin DW: Alcohol in suicides and homicides. *Q J Stud Alcohol* 34:144–156, 1973.

72. Gross MM, Hasty JM: Sleep disturbances in alcoholism, in Tarter RE, Sugarman AA (eds): *Alcoholism: Interdisciplinary Approaches to an Enduring Problem.* Reading, Mass, Addison-Wesley, 1976, pp 257–308.

This chapter is a revision by permission of the author and publisher of *The Effects of Combined Alcohol-Drug Abuse on Human Behavior,* originally published in *Drug Abuse and Alcoholism Newsletter* 2:1–13, 1979.

6 Combined Treatment of Alcoholism and Drug Addiction

Donald J. Ottenberg
Jerome F.X. Carroll

HISTORY AND DEVELOPMENT OF COMBINED TREATMENT

Some Definitions

Combined treatment is the treatment in the same program of persons considered to be alcoholics and others considered to be drug addicts or drug abusers, as well as persons who have abused both drugs and alcohol. All types of substance abusers are treated together by the same staff using the same treatment procedures. Stated otherwise, although treatment regimens may vary to meet differing individual clinical problems, treatment does not differ simply on the basis of the diagnosis of alcoholism or drug addiction.

The essential feature of combined treatment is that persons with different substance-abuse histories and preferences are fully integrated into a single treatment program and process. Even though a treatment agency provides services to both drug addicts and alcoholics, it is not conducting combined treatment unless these two types of patients are treated in the same place, at the same time, by the same methods, and by the same staff.

The concept of combined treatment is most meaningful when applied in residential treatment settings, where persons of various backgrounds and substance-abuse experiences mingle in everyday living situations, as well as during therapy per se. Combined treatment also is applicable in outpatient locations, although the impact on patients and staff and the issues that arise are different from the experience in the live-in environment.

The Generic Concept of Addictions

The generic concept of addictions (or substance-abuse problems) is a term that has been used to convey the idea that all substance-abuse problems are part of the same problem, even though choice of substance varies from person to person, and manifestations resulting from substance abuse differ from drug to drug. That one person is addicted to heroin and another to alcohol is less important than the fact that both are abusing some chemical substance capable of leading to dependence and addiction.

The choice of a particular substance and consequent development of dependence to that substance (or group of substances) is in large part accidental. Who becomes an alcoholic and who a narcotic addict or psychedelic substance abuser appears to depend on chance factors of time, place, and circumstance, as much as on individual constitutional or psychogenic factors. Where and when one was born, into what race, and with what substances available and being used by family members and/or one's peers are factors that together play a primary role in determining whether one will become substance-dependent and to which substances.

This does not deny that some persons experiment with a variety of substances until they find what they like best, or that others practice self-medication, switching back and forth among several types of substances, depending on the kind of effect that is desired at the particular time. The point being stressed is that social, cultural, and family variables are powerful determinants in contemporary life, and these factors are not specific to particular substances.

Evaluating Differences from the Generic Perspective

The special characteristics of various addictions or substance dependencies that makes them distinguishable from one another develop as a consequence of addiction to specific substances. Thus, among the various substance dependencies, withdrawal states differ, if they occur at all. Tolerance develops at different rates and to different degrees, if at all. Cross-tolerance occurs with different groups of substances, if at all. These differences, which some observers point to as evidence of specificity of various addictions, are differences resulting from the specific pharmacologic and toxic properties of various chemicals. The existence of different qualities and effects of different drugs, however, does not negate the rationale for considering all types of substance abuse and dependence as varieties within a single class or genus.

These "specific" differences that at certain times separate the alcoholic from the LSD abuser and the heroin addict from the glue sniffer are of utmost importance for the prompt and proper treatment of acute syndromes that occur in substance-abusing persons. In this phase of treatment, one is dealing with specific effects of drugs, calling for specific treatments. This is a medical phase of treatment in hospitals, medical clinics, emergency rooms, and physicians' offices where persons go for help or are taken when suffering from acute withdrawal symptoms or from the toxic or complicating effects of drugs. Once the acute, medically supervised phase of treatment is over, however, these specific, distinguishing variables dissipate and henceforth have relatively little importance in the treatment of the underlying addiction or substance dependence.

A generic concept has no difficulty accommodating variations that result from the interaction of an individual person and a specific substance or substances. These are specific differences that do tend to separate different types of substance dependence from one another at certain times. What is not acceptable are popularly held perceptions that imply either absolute or greater differences than similarities between alcoholics and other substance-abusing groups for which little valid scientific evidence exists to support such claims.

Most of the perceived differences that distinguish one type of substance abuser from another have to do with such factors as age, race, generational lifestyle and values, language, degree of conformity to conventional modes of behavior, and so forth. None of these variables, however, is specific to any substance of abuse. Moreover, when researchers either "match" or statistically control for the effects of these variables in making comparisons of alcoholics and other substance abusers, far more similarities than differences are noted.[1-4]

The statement, "You can't treat alcoholics and heroin addicts together, they're too different," conceivably might be true. But if it were true, it would be because of differences of age, culture, and lifestyles, and attitudinal blocks of patients and staff about these factors—not because of any intrinsic difference between two entities, alcoholism and heroin addiction. Clinical experience with various kinds of addicted persons informs one that in looks, thoughts, and behavior, young alcoholics are much closer to heroin addicts of the same age group than they are to older alcoholics.

Combined Treatment and the Generic Concept Are Not Equivalent

Combined treatment and the generic concept are related, of course, but they are not the same thing. Generic conceptualization takes into account the nature of addiction and its etiology, including social, cultural, economic, and political elements, as well as intrinsic biologic variables. The generic concept provides arguments to support the idea of combined treatment, while encompassing issues that range far beyond the province of treatment. Combined treatment is, after all, simply treating various kinds of substance abusers together and in the same way, with or without an ideological rationale for doing so.

Traditional Approaches to Treatment

An ironic footnote to the federal statutes that governed the treatment of narcotic addicts for many years, until the current era began in 1965, is that those laws made it almost impossible for the great majority of physicians to acquire experience in the treatment of narcotic addiction. Heroin addicts were treated in the federal hospitals at Lexington, Kentucky and Fort Worth, Texas. A small number of narcotic addicts, mostly medical professionals, were treated in private psychiatric facilities, with permission of the Drug Enforcement Agency.*

Most medical students had no opportunity to study narcotic addicts, except through textbooks or lectures, and very few physicians came face to face with narcotic users. Questions about the feasibility and desirability of combined treatment were meaningless, and a generic concept of addictions was thought about only in theoretical terms by a small group of physicians, pharmacologists, physiologists, pathologists, psychologists, and sociologists who had a special interest in the subject.

Alcoholics were treated by medical practitioners in hospitals, outpatient clinics, and private offices for the medical complications of drinking or an acute withdrawal state. For treatment of the underlying

*Formerly the Federal Bureau of Narcotics and Dangerous Drugs.

illness, when there was such treatment, the alcoholic was referred either to a psychiatrist or to Alcoholics Anonymous. Most psychiatrists at the time treated the alcoholic as they would treat any other person with a psychiatric problem. The addiction to alcohol was viewed as a manifestation of neurosis or character disorder. When the basic emotional or psychological disorder cleared, the alcohol problem was expected to be resolved along with other symptoms.

In the 1930s, 1940s, and 1950s, there were a few alcohol "specialists" scattered about the country. They were physicians, mostly psychiatrists, internists, or generalists, who, for various reasons, including personal involvement with alcohol, had developed a special interest in this problem. These physicians rarely saw a narcotic addict.

Physicians—and others—who saw significant numbers of alcoholics came to realize that many alcoholics also misused other drugs, particularly sedatives, and, when they became available, tranquilizers. This fact, which lends support to a generic view of the problem, has been appreciated only slowly by the medical profession at large. Even today some physicians do not understand how easily, and perilously, any alcoholic can be converted to chlordiazepoxide or diazepam addiction.

An important and enlarging source of help to alcoholics in the period from the mid-thirties to the early sixties was Alcoholics Anonymous (AA). Many alcoholics found their way to AA through the help of family, friends, and other AA members, and quite a few physicians began to use AA, along with Alanon and Alateen, as a primary source of ongoing help for alcoholic patients and their families.

Status of Substance Abuse Treatment at the Beginning of the Current Era

We date the current era from the time of passage of the Narcotic Addict Rehabilitation Act (NARA) in 1965. The general purpose of this law was to decentralize the management of narcotic addiction and to provide mechanisms and funds with which to offer services to narcotic addicts in their own communities, both before and after treatment in the federal treatment centers. Later, the provisions also allowed for treatment to take place in the local community. An important effect of NARA was to awaken interest in this problem in the professional health care community and to create means by which various health care workers could have direct experience with addicted persons. This was the beginning of a burgeoning field of health care work that attracted thousands of practitioners over the next decade.

After NARA—and related drug and alcohol federal legislation that followed—health workers of various kinds joined the field. The list included physicians, psychiatrists, psychologists, social workers, counselors of various types, rehabilitation specialists, teachers, and group and individual "therapists." In the category of "therapists," all the current schools of therapy were represented: gestalt, psychodrama, transactional analysis, encounter groups, reality therapy, and others.

Some of the earliest arrivals were persons who had recovered from addiction and now sought to follow this work as a new career. In many places, the recovered staff, both alcoholics and drug addicts, though usually not in the same program, laid the groundwork on which many traditional professionals arriving later were able to build programs and careers.

The recovered alcoholic and the ex-addict staff members brought with them a strong separatist bias, which typically was also shared by their nonrecovered, professional co-workers. Each group, being unfamiliar with the addictive background and lifestyle of the other group, tended to see different substance-abuse problems as quite distinct. It did not occur to most recovered alcoholics and ex-addicts that they might be similar in more ways, or in more important ways, than they were different. Perceiving themselves as different, they believed the treatments required for the two types of addiction had to be different.

This wide and nearly unbridgeable gap between alcoholics and drug addicts was strongly supported by AA in its doctrinal expressions and by the National Council on Alcoholism (NCA). The Committee on Professional Relations of the General Service Board of AA,[5] for example, wrote, "AA is concerned with one thing only—the recovery and continued sobriety of those individuals who come to the Fellowship for help with their drinking problems. AA membership is confined to alcoholics. We welcome opportunities to share AA experience with those who would like to develop self-help programs for nonalcoholic addicts—using AA methods, perhaps, but using the *experience* of the nonalcoholic drug addict during addiction *and* recovery" (p 2).

Pittman's[6] "The Rush To Combine," captured the spirit of this separatist view when in 1967 he expressed concern that the fields of alcoholism and other drug problems would merge and, in doing so, damage the opportunities to collect necessary data and carry out needed research. By December 1974, however, Pittman had changed his mind and stated so publicly in San Francisco at the North American Congress on Alcohol and Drug Problems.

The creation of the National Institute on Alcohol Abuse and Alcoholism (NIAAA) in 1971 and the National Institute on Drug Abuse (NIDA) in 1973 removed these two fields of work from the National Institute of Mental Health (NIMH) where they had resided since 1949. The federal legislation, and the large sums of "new" money distributed as a result of it, led to the creation of many new programs and new jobs throughout the country. Most of the workers were employed either by an "alcohol agency" or a "drug agency." Very few worked with both problems.

Even physicians working in these fields tended to join one camp or the other, contrary to the traditional philosophy of most physicians that they treat the person, not the disease. The American Medical Society on Alcoholism, born in 1967 as an outgrowth of the New York City Medical Society on Alcoholism, chose to use "alcoholism" rather than "addictions" in its title and later affiliated with the National Council on Alcoholism in 1973 as its medical arm.

To our knowledge only one effort at combined treatment had been attempted in this country before 1968, when Eagleville began to admit narcotic and other drug addicts to its existing alcoholism program. The Silver Hill Foundation[7] in Connecticut experimented with admitting adolescent drug abusers into its traditional alcoholism treatment program. The results were hectic and very negative. Neumann and Tamerin[7] concluded from this experience that combined treatment was not feasible or effective, and they abandoned the program in their institution.

Factors Leading Toward Combined Treatment

In an article written as a counterpoint to Pittman's,[6] Ottenberg[8] noted that:

> Forces beyond anyone's control account for much of the recent preoccupation with the possibilities of combined treatment for alcoholics and other substance abusers and a generic, rather than substance-specific, conceptualization of addictive problems. The trend of thinking toward combined approaches stems directly from growing awareness of the increasing prevalence of multiple substance abuse, the shifting from one substance of abuse to another, and the frequent failure of treatment for substance abuse because of "escape" into other substances of abuse. A prime example of this latter phenomenon is the widely reported abuse of alcohol as a cause of failure in the methadone treatment of heroin addicts.
>
> Furthermore, single substance abusers are becoming hard to find. They are mostly confined to older age groups. Among the young, multiple substance abuse is typical, and alcohol is just one more drug abused with many others. (p 280)

We think this statement is as true today as it was in 1977. Multiple substance abuse remains the typical pattern of substance abuse, particularly among the young who are most likely to abuse both alcohol and other drugs.[9-14] Extensive studies and numerous reports of polydrug abuse, which we call multiple substance abuse, attest to its prevalence and importance.[15-17]

The National Drug/Alcohol Collaborative Project (NDACP), in which ten federally funded programs around the country participated, also reported the significant occurrence of multiple substance abuse and illustrated some of its patterns and consequences.[18,19] We feel certain that by this date most clinical workers in the drug and alcohol field have observed individuals who shift from one substance of abuse to another, as well as those who abuse several substances at the same time. The distinction, moreover, between those who abuse alcohol and those who abuse other drugs is rapidly diminishing in importance.

We place special emphasis on the transferring of dependence from one substance to another that occurs during treatment.[20] The professional literature contains many reports of alcohol abuse—and alcoholism—as a complicating problem in the methadone maintenance treatment of narcotic addicts.[20-23] In fact, some observers cite alcohol abuse as the most frequent and most serious cause of failure in methadone treatment.[24-26]

These facts take on added significance if the staff of a treatment program is not prepared to meet these problems therapeutically or, much worse, if their ignorance of this aspect of addictions allows them to do what may be detrimental to persons in their care. An example would be the failure of a counselor or therapist in a "drug program" to realize that a person in treatment was showing evidence of problem drinking and was moving rapidly toward alcohol addiction.

Until recently (and perhaps even now) one could find treatment programs for heroin addicts that ignored the hazards of drinking for their patients in treatment, and perhaps even fostered the abuse of alcohol by condoning or ignoring excessive drinking by staff and patients.[27] Similarly, many alcohol treatment centers failed to inquire about the abuse of "medications," pills, and marijuana. Worse yet, sedatives and tranquilizers were often prescribed routinely for alcoholics and alcohol abusers with little regard for their potential for abuse or threat to health.

We do not wish to accuse or condemn, but we do feel obligated to stress the urgent need of the field to prepare its workers adequately for the clinical responsibilities they will carry. This alone is good reason to take a generic approach in the training of staff, whether they intend to work with drug addicts or alcoholics.

Demands for broader and more generic kinds of training today are coming from front-line staff. They appreciate, probably better than those in supervisory and administrative echelons, that problems fitted comfortably into distinct categories for legislative or bureaucratic reasons may not stay separated in this way clinically. Staff in all programs need training and experience that enables them to understand, recognize, and properly manage problems related to various patterns of substance abuse. Anything less than this today shortchanges staff and imperils their competence for the work they do.

We are not talking here of training staff to conduct combined treatment. We are advocating training and experience that are sufficiently broad to cover the common varieties of addictive problems that are encountered today, especially patterns involving the abuse of alcohol and other drugs. This can happen only if training programs, many of which are supported by federal and state funds, will allow curricula to cross established and usually well-guarded territorial boundaries.

Evolution of the Eagleville Combined Treatment Program

Eagleville began its treatment program for alcoholics in 1966 after a long career as a private, not-for-profit institution devoted to the treatment of tuberculosis.[28] Essentially the same board of directors who had had ultimate responsibility for Eagleville as a tuberculosis hospital made the decision to shift to the new purpose of study, training, and treatment of alcoholism when the hospital's beds no longer were needed for patients with tuberculosis.

The new program was organized by a small staff of physicians (one of whom was a psychiatrist), nurses, social workers, psychologists, rehabilitation counselors, therapists (some of whom were recovered alcoholics), and support staff. The early program was organized on a fairly conventional medical model for a chronic disease hospital.

Only after narcotic addicts and a few drug abusers of other kinds were added to the patient population beginning in late 1968 and early 1969, did the character of Eagleville begin to shift toward a therapeutic community-type of organization. In fact, Eagleville never ceased being a hospital and has always had credentials and funding as a hospital, yet its administration has included many features of a therapeutic community.

Patients, called residents, are assigned to groups, and groups with their staff are organized into teams. The entire body of residents and staff meets five mornings of the week for "community meeting," where issues of interest to the whole community are discussed and various reports are made. There is a continuous effort to have all members of the community, residents, and staff of all kinds and levels, share responsibility for maintaining a safe and secure drug-free environment.

There are no guards or police, so that maintenance of an environment conducive to therapeutic effectiveness and personal growth is more a matter of the combined conscience and community ethos than of surveillance or mechanical barriers. There are rules bearing on conduct and responsibility, the most important of which have to do with "using" substances or bringing drugs or alcohol on the grounds. Of equal or greater importance are strict rules banning the use of violence. Infractions of rules provide opportunities for therpeutic interventions on an individual, group, or community-wide basis.

Sometimes residents are discharged or required to leave the Eagleville campus for a specified period, depending on the nature of the incident and the status of the rules at the time. Since Eagleville makes a greater effort to govern itself in an open and consensual fashion than does the usual hospital, rules and penalties shift back and forth on the axis of stringency.

At Eagleville's 72-acre main campus are housed the 126-bed Inpatient, or Hospital Program; the Candidate, or Transitional Program, with 70 beds; and various special services, such as research, training, and Administration. The Inpatient Program is an intensive initial phase of diagnostic study and treatment lasting usually about six to seven weeks.

The Candidate Program is geared more to preparing people for the return to the outside community. Its members work half-time at paid jobs in the hospital system and spend the other time in therapeutic and rehabilitative activities. This program extends over four to six months.

Eagleville also operates a number of specialized programs, most of which are located in surrounding communities away from the Eagleville campus. These programs include a Family House, where substance-abusing women can live with their young children while undergoing treatment; halfway houses, one for men and one for women; the Montgomery County (Pennsylvania) Community Day School, where young people of high school age who have been in trouble in school and in the juvenile courts attend what is both school, with official academic credits, and a day therapy program organized along therapeutic community lines. The point to be emphasized is that all of these programs—on and off the Eagleville campus—are combined programs that accept alcoholics, drug-dependent persons, and multiple substance abusers without distinction.

In this brief review we cannot offer an in-depth discussion of how the Eagleville combined therapy programs were conducted and the effect of adding drug-addicted persons, principally heroin addicts, to the existing patient population of alcoholics. These accounts are published elsewhere.[29-31] What we wish to do here is to indicate briefly some of the more significant aspects of combined programming and its effects on a treatment program.

The first point to be made is that alcoholics and drug addicts, to the extent that they can be separated into two categories, differ in age, lifestyle, and behavior, both around substance use and in other ways. Each group tends to view the other with suspicion, contempt, and hostility. The drug abusers are seen as violent, dangerous, and criminal by the alcoholics, who, in turn, are viewed as old, stupid, and contemptible. These stereotypes are no different from those held by the general population of similar age and educational and socioeconomic status. Even staff who have not had previous opportunities to work with both groups are likely to share these prejudices.

These perceptions of a group of persons different from oneself, expressed in scornful and fearful stereotypes, provide an excellent therapeutic opportunity, if the staff is capable of constructively responding to these biases. In reality the alcoholic residents and drug-dependent residents have much more in common than they are aware of, and the similarities that soon become evident are far more important than more superficial differences.[1,3,4,29,30]

The introduction of drug-using persons into a population of alcoholics—or the other way around—almost certainly will result in a more heterogeneous population, with greater variety of race, culture, economic, educational, vocational, and social background, and, not to be ignored, experience in criminal behavior and legal processes related to it. This comes closer to the ''melting pot'' of American tradition. It results in noise, misunderstanding, misapprehension, and interpersonal difficulty. At the same time, however, it provides opportunities for persons to understand and appreciate themselves and one another in ways that can benefit them profoundly. Much depends on the staff and their readiness and competence to use the situation productively.

Many older alcoholics look at a younger drug addict and, despite the difference in substance used, see themselves 15 or 20 years earlier. In expressing this identity, which one addicted person can do so vividly for another, the differences of substance preference and lifestyle related to the use of a particular substance fade into relative insignificance in the presence of the core problem of being substance-dependent that links the two persons.

Age and cultural differences also offer opportunities to work out parent-child conflicts that are often initiated or complicated by generational incompatibilities. For some fathers and mothers, this is the first real understanding they have of their own alienated children and the values their behavior expresses. In the other direction, some young persons gain a new appreciation of parents and the difficulties faced by them within the context of a different set of beliefs and norms.

Inherent in all the interactions is a focus on the individual's situation and needs and an attempt to find effective, appropriate responses to those needs. Labeling a person ''alcoholic'' or ''drug addict''—or any other ready-made diagnostic tag—is not seen as a sufficient basis on which to provide treatment and rehabilitative care. In going past superficial distinctions in the attempt to grasp and administer to essential issues of individual personality, character, behavior, and life situation, the fact that a person is involved with this or that substance, or group of substances, quickly ceases to be of major importance.[32,33] Indeed, after the first day or two at Eagleville, very little time is spent dwelling on the specific substance or substances whose abuse was the stated reason for admittance.

Recently, Eagleville's generic approach to substance-abuse rehabilitation was codified in terms of a two-phased, sequential treatment goal model developed by the second author. Regardless of what substance(s) the person has abused, the sequence of treatment pertains, even though the treatment objectives are always suited to the particular individual's needs and circumstances. The sequence is as follows:

Phase I

Detoxification and continuing detoxification
Restoration of physical health
Restoration of emotional stability
Motivation for treatment
Aftercare planning (acknowledgement of the need for continuing aftercare treatment)

Phase II

(Early)
Confronting pathological defenses
Facilitating catharsis and insight
Enhancing self-esteem
Aftercare planning (developing specific and concrete aftercare treatment plans)

(Late)
Aftercare planning (preparation for discharge and implementation of aftercare plans)

The treatment objective for staff is to bring each patient to a "minimally effective level of functioning" for each goal before proceeding to the next stage of treatment. The pace of treatment varies from patient to patient, but usually Phase I goals are the focus of treatment for the first seven to ten days. Early Phase II goals receive major attention usually from the third to fifth week, while the late Phase II goal typically is the center of attention during the last two weeks of treatment.

Some behavioral differences need to be taken into account and, providing staff is willing to work with heterogeneous groups, they readily become knowledgeable and effective in relating to these different lifestyles, cultures, values, and behaviors.[34] Persons primarily involved with heroin, cocaine, and cannabis are not only younger, as a group, than those who have preferred alcohol, but they also have experience in a different street life that includes petty crime, shoplifting, burglary, and the like. They probably have been in prison more recently than the alcoholics, and they are more likely to find the opportunity to join forces with one another in conspiracies to acquire drugs or breach other important rules of the treatment community. They make more trouble (of this sort) than the alcoholics, who, despite their addiction and all the failure it implies, are a more mature and, in conventional terms, a more stable group. The core problem of the alcoholics is more likely to be depression and a sense of hopelessness and defeat.

Each group can help the other. The older, more reserved persons are a buffer and control on the acting-out tendencies and disruptive behavior of the younger group. The younger persons, in turn, create in the community an energizing and revivifying atmosphere which can be an antidote to resignation and despair. This happens spontaneously but, of course, not without influence by the staff who must guide interactions into therapeutic channels.

EFFECT OF COMBINING ON THE TREATMENT ENVIRONMENT

When a younger group of residents comes into an existing program or treatment community, inevitably they bring with them additional noise and disruption of established routines. The community must develop means of meeting the new disorder, without being so suppressive as to deny the younger group its rightful place and opportunity. This process is mediated in innumerable confrontations, small and large, each of which helps to define the community's standards and expectations. In this process residents are exposed to instructional experience not easily provided in more formal ways. People learn to give and take, to make demands, and to yield to the rights and needs of others. For many persons these are important encounters that reveal oneself and provide fresh insight into one's own needs and responses. Even learning to yield, and endure frustration, can be an important achievement for persons not only chemically dependent but emotionally immature.

A discernible effect of having a broader and more varied population in the treatment program is the challenge it presents to each person's preconceptions and prejudices concerning groups or types of people different from oneself. Enlarging the spectrum of substance-abuse problems almost certainly will increase the variety of persons in the community. The mix of the population will vary from one locality to another; however, in all likelihood there will be a greater range of economic, racial, cultural, social, and educational backgrounds.

At first the increased variability causes discomfort, for people naturally tend to feel more secure among those they perceive to be like themselves. In the long run, the "melting pot" effect can be one of the most potent forces pressing toward personal growth and needed change. One's prejudices, left undisturbed, continue to be a limiting and, at times, destructive internal force. They express a distorted view of reality, including the self, nurtured, as they are, by a compost of fear, self-doubt, and failure of self-acceptance. Part of the work of recovery from addiction takes place in these very precincts, and having to deal with feelings engendered by a more heterogeneous population is not a diversion from primary therapeutic purposes.

These ideas are consistent with a concept of treatment that expects, and even requires, that profound changes take place within the individual in the process of recovery. For a young black man from one of our larger cities, whose entire perception of life is as a series of insults and oppressions, it is virtually impossible for recovery to be achieved without working through, in some durable way, accumulated anger and resentment.

For a white person, or someone more affluent or "privileged," self-esteem may depend more than is realized on unfounded hatred or contempt for people of another race or social class. Here there can be, at best, a shaky sobriety in danger of collapse, unless one deals with fundamental issues of self-regard and relation to others.

We understand that these views differ from the philosophy of some treatment programs where the aim appears to be to help people achieve sobriety with as little disturbance of the individual as possible. In these programs a homogeneous population is an asset, and shifting to a combined treatment agenda could cause great disturbance unless the philosophy of treatment also changes.

ORGANIZING AND CONDUCTING COMBINED TREATMENT

Staff—at every level—constitute the most important determinant of success in combined treatment. First, they must be committed to the idea of combined programming and sincerely be willing to practice it.

Second, they must be competent to provide effective treatment in a combined treatment environment.

When Eagleville made the decision to admit narcotic and other drug addicts into its existing program for alcoholics, the policy change was treated as an important institutional decision. The question was widely discussed within the Eagleville family and ultimately was decided by clinical staff, administrative staff, and the board. In the course of the decision process, some of the staff visited drug programs, attended lectures and other educational meetings, and read the drug "literature."

When the decision was finally made, it was agreed to make the move quietly, that is, without publicity or public notice. It was clear, however, that the policy change had the full, legitimate stamp of approval from Eagleville as an institution. It was *not* one staff member, or a few, being "tolerated" as he tried "something different." All of Eagleville was responsible for the innovative step, even though only one narcotic addict was admitted at first, and very few staff had direct clinical responsibility.

We stress the comprehensive approach taken in making the decision, and the full staff and board involvement in the process, because we believe that anything less may fail to uncover some of the deep feelings and fixed attitudes that exist in any group of persons working in a professional setting. In neglecting to permit full and open expression of ideas and feelings, the seeds of failure of the program may remain, buried at the moment, but capable of growing in the future into a destructive force. In direct language, we believe that half-hearted attempts in combined treatment are doomed to fail.

That genuine support must come from every part of the organization does not mean that everyone will be of the same mind. At Eagleville, a couple of staff members decided to leave and, over a few months time, so did two or three veteran members of the board. Those who stayed held various opinions about the wisdom and feasibility of the decision. Everyone remaining, however, was fully aware of what was being done and was willing at least to try to make it work.

Selection and Training of Staff

Staff who have had good training and experience and who have demonstrated ability to work effectively with a particular addicted population are almost certainly capable of working well with other addicted populations. The exceptions will be persons who lack the interest or versatility to take on this new challenge. For example, a recovered alcoholic staff member may find it difficult to identify with the younger group of drug abusers and cannot feel comfortable working therapeutically with this group. The same problem may occur for the recovered drug addict who cannot identify with or relate to older alcoholics. Where the general atmosphere of the program encourages cross-fertilization of ideas and experience and an effective network of mutual support exists among staff, the staff find little difficulty in establishing helpful therapeutic relationships, regardless of age, cultural background, and drug-taking experience.

The presence of both recovered alcoholics and recovered drug addicts on the staff is of inestimable value in a combined program. Their experience and example offer the best antidote to misguided preconceptions and barriers based on mythical stereotypes. Nothing extraordinary need be demanded of the recovered staff members. It is sufficient that they participate in day-to-day work relationships and, like everyone else, have opportunities to know other staff members and be known in return. Later, as all staff become sufficiently imbued with the generic idea to be able to defend it, the recovered staff will seem much less special.

Training: A Key Component of Success

The quality of training of staff for combined treatment is of critical importance. Obviously training must be broad enough to encompass familiarity with various kinds of substances and the manner and consequences of their use, including patterns of multiple-substance abuse.

Along with didactic instruction exploring similarities and differences of various drugs and drug problems, there should also be opportunities for trainees to express conflicting ideas and feelings on various issues. These include reactions to various types of substance abuse and persons who abuse those substances, lifestyles and behavior of abusers, feelings about working with such persons, and views about personal use of various substances—including their use by other staff.

Allowing trainees to explore their ideas and attitudes and to express their feelings is also essential. In this regard, special affective training techniques may be needed to uncover hidden prejudice, hostility, and fear. Certainly, an ambiance of trust will be needed in the training environment. Ultimately, the competence of staff to work in a combined setting may be influenced more by this part of training than by the more conventional instruction concerning epidemiologic, pharmacologic, psychologic, sociologic, and medical aspects of various types of substance abuse.

Training also should include opportunities for direct interactions between trainees and persons who have abused various substances.

Recovered staff in particular need an opportunity to interact with persons who have had substance-abuse histories different from their own.

Whenever possible, trainees should visit various kinds of programs organized for substance abusers or addicts of various types. Trainees who have had work experience with one type of substance abuser certainly should be afforded an opportunity to see programs for other types of abusers. The list of possibilities from which to choose is large. We believe that every trainee should have a first-hand look at methadone, and possibly other chemotherapeutic approaches, as well as drug-free programs for heroin addicts, both residential and nonresidential. Other programs that should be visited are alcohol rehabilitation programs, and medical and nonmedical detoxification programs for both alcoholics and other substance abusers. If actual visits are impossible to arrange, use of video tapes and outside consultants or trainers could serve as acceptable substitutes.

Visits to both Alcoholics Anonymous (AA) and Narcotics Anonymous (NA) should be part of every training program for clinical staff. Open AA and NA meetings are available for this purpose in almost every community. The best way to arrange for these experiences is through the local intergroup office of AA or NA, if one exists. Usually a member of AA or NA will be pleased to act as host for one or two visitors, thus providing invaluable instruction and insight along with hospitality.

Through presentations and visits the trainees should become aware of the community's organized response in the addiction and related fields at federal, state, and local levels. All of the training should be integrated in such a way as to help trainees understand addictive problems and programs as part of a generic concept. This way of viewing information and questions should become fully incorporated by each trainee, since it will be the conceptual base on which work in the field will rest.

Some Programmatic Considerations

Again we caution not to allow the phrase "combined treatment" to obscure the fact that treatment is planned and executed to meet individual needs, just as it is in any program. Some parts of treatment are shared with other patients, namely those who have similar therapeutic indications, while some are not. With respect to space, living quarters may have to be shared. In this case, the needs and rights of all have to be considered and looked after. This is no more or less proper and necessary than it would be if all the residents were diabetics or cardiac patients.

If a program cannot assure an environment that is suitable for all the residents who must share it, failure, when it inevitably comes, should not be ascribed to combined treatment. Rather, it should be attributed to the fact that the program was not properly organized and conducted.

We offer an example. A hospital-based addiction program treating narcotic addicts and young substance abusers as well as older alcoholics was attempting to take a combined-treatment approach. A small detoxification unit was created within a larger area used by both the drug addicts and the alcoholics.

Because of the physical location of the detoxification unit, much of the noise of the routine activities of the larger area carried into the detoxification unit. Soon the unit had to be moved, because it was not possible to provide proper treatment there. Toxic, severely ill alcoholics experiencing withdrawal syndromes of psychomotor agitation could not tolerate loud radios, guitar playing, and the commotion of young people interacting with one another. It also had become evident that medications left unattended at the bedside of bedridden patients were being stolen. These procedures simply were not appropriate.

Similar questions arise in other phases of treatment and call for reasonable solutions. For a group of persons of varying ages, tastes, and tolerances to live together peacefully, certain rules with fair give-and-take are necessary. A situation of this kind requires set places and times for noise and for quiet, just as it would in a house shared by a family. We mention these requirements, as obvious as they seem, because we believe that some attempts to combine treatment have foundered for having failed to manage these simple issues.

ADVANTAGES OF COMBINED TREATMENT

In considering some advantages and disadvantages of combined treatment, we are aware that no single approach will satisfy everyone's needs, and some feature advantageous to one person may actually be against the interest of another. This again requires that attention be focused on the individual and his or her unique set of problems, rather than on the more abstract question of involvement with a particular substance.

The generic treatment philosophy recognizes that all forms of substance abuse have many dynamics in common, especially a negative self-image. Thus, all substances of abuse are considered to be potentially threatening to the recovering substance abuser. This perspective can create a powerful ethos against the tendency of many patients to shift from one substance of abuse to another, particularly under the stress of treatment, and assures continuity of care should a shift to a different type of abuse occur.

The generic perspective also is helpful to persons in the process of making the difficult interior decisions required to interrupt an addictive pattern of living, in that it offsets the natural tendency of addicted persons to reify the addiction and view it as something external that "happens" to them. The person, regardless of the substance(s) abused, is pushed toward acceptance of responsibility both for being addicted and for taking the steps necessary for recovery. While the objectives of treatment include the discontinuance of substance abuse and maintenance of sobriety, the more fundamental goal is to help the person change in ways that will open life to greater fulfillment.

We think that a combined approach accurately reflects the drug scene of today, where it is not unusual to see families in which a drug-abusing son or daughter is the product of a household in which the mother or father (sometimes both) is an alcoholic. Approaching these problems in two generations as related and in many ways similar can be helpful in managing the individuals involved and the family as a whole.

Combined treatment also offers a number of advantages to the staff, the benefits of which filter down to patients. Staff have opportunities for broader experience and training in the addictions field. This is valuable for everyday professional work and also opens wider career opportunities. Combined programs may also be more useful than segregated programs to the communities in which they reside, providing, as they do, more comprehensive resources of information, education, prevention, and community program skills.

Disadvantages of Combined Treatment

Disadvantages fall into three categories: problems created for the patients, the staff, and the program as a whole.

Some persons do not wish to submit to treatment that aims at helping them by pressing for fundamental changes in personality and lifestyle, or that sees the achievement of sobriety as only one objective within more important therapeutic goals. What such persons want—and can get in many programs—consists of education about what addiction is and how it affects one, support in becoming motivated to stop drinking or using drugs, and instruction in ways to avoid the "people, places, and things" associated with substance abuse.

These programs offer much sound advice about where to find

helpful and durable support outside the program. For many persons, AA will be the principal source of continuing support, and an understanding of AA policies and methods and numerous introductions to the fellowship of AA may be built into the rehabilitation program.

We are contrasting this kind of program with what we have described as a generic approach, illustrated by our combined treatment program. Our purpose is not to denigrate the value or effectiveness of other programs but to emphasize the difference between approaches that undertake to assist individuals to bring about significant changes in personality, self-concept, and behavioral patterns and other less penetrating approaches. We believe that the generic approach has difficulty avoiding penetration of the individual in the sense we have been discussing, and a person who does not wish to be exposed to this kind of challenge to self-concept and established patterns of thought and behavior is probably better off in some other kind of program. Some older persons, especially more passive and perhaps timid alcoholics, may not do well in an atmosphere that becomes too charged with excitement and conflict from time to time. It may be too frightening.

The very presence of many persons of different colors, values, and cultural origins may be upsetting to some people. If they have the will and courage to withstand the discomfort, the ultimate gains from the experience may be very valuable; for some, a unique opportunity in their lives. Exposure of self, with one's prejudices and distorted ideas in evidence, can be embarrassing and even cause anguish. It is not what some people came to treatment for, and some are not willing to pay the price, no matter how worthwhile the gains might prove to be.

Some people want to feel at ease at almost any cost, and this usually means to be among people of one's own kind. In our country, feeling at ease is related more to social class than to other variables, such as race, age, or preference in substance use. Pointing out to prospective patients that "feeling comfortable" is not the most important consideration in one's choice of a treatment program is not very convincing to many people. In our experience, the higher a person's economic status, the more likely the person will choose a treatment environment that feels "comfortable" and "safe."

For staff, working in a combined program offers special advantages, as we have indicated, but the personal cost in emotional "wear and tear" can be high for some. This is especially true in programs with considerable heterogeneity of staff as well as patient population. If some staff are traditional professionals with the usual academic training and credentials, and other staff are nontraditional professionals who lack academic degrees, there is inevitably a continuing disequilibrium, a tension created by the encounter of persons from different worlds, with different opportunities and rewards, and with different perceptions of themselves and one another.

Like any other focus of conflict in a mixed community, this can be an opportunity for growth and improvement for all concerned (including the patients), but the effort requires large quantities of time, energy, and emotional interaction. Added to the many frustrations and difficulties associated with working with addicted persons, this can be an additional factor in the multivariable causation of staff burnout or battle fatigue, a problem of significant proportions in the addictions field.[35] Many efforts are made to offset these hazards; however, space does not permit a discussion of this here.

A staff that includes recovered alcoholics, recovered drug addicts, and persons who have not been substance-dependent faces particular issues revolving around drinking and drug use by staff members. Most recovered alcoholics avoid drinking totally, feeling that unqualified abstinence is necessary for continued sobriety and stability. Many recovered drug addicts, particularly heroin ex-addicts, drink "socially." They believe that alcohol is not a problem for them and, based on the fact that many drink without getting into difficulty, they appear to be correct in this view.

It is evident, however, that some former heroin addicts cannot use alcohol without moving toward abuse and addiction, and suffering harmful consequences. Clearly, these individuals must not drink. What proportion of recovered heroin addicts belong in this category is not known. Some of our data show that an individual's experience with alcohol and its effects in the period before heroin addiction ensued, offer useful indicators of how alcohol will affect the person after heroin addiction has been overcome.[20]

In practical terms, whether a recovered heroin addict can drink safely or not is an empirical question, answered, if the individual wants an answer, by experimentation. That some "recovered" staff drink and others do not creates tensions. The fact that some staff, who themselves were once addicted, are now drinking deprives the program of a solid front of opinion and commitment that could be a powerful influence in helping patients to become motivated to live drug-free.

Marijuana smoking and use of other forms of cannabis poses a similar problem. Here, the substance may be less of a threat to the recovered alcoholic staff member than to the recovered drug addict. This, too, remains undetermined. The fact that marijuana use continues to be an illegal act in most parts of the United States makes it easy for any drug program to adopt an official policy against its use by staff. This does not alter the fact that some staff in some programs do use cannabis "recreationally," and the more mixed the staff in a program, the more difficult this issue is likely to be.

Social drinking and "recreational," nonabusive use of marijuana by staff who have never been addicted also can become an issue of contention. The greater in numbers and more powerful the recovered staff who insist that total and permanent abstinence is the "only" way, the more likely will conflict emerge among staff around such substance use.

These problems among staff bear on the patients in two direct ways. They are witness to what transpires with staff who are viewed as role models. Conflict and confusion among staff is transmitted to the patients. The problem also affects patients by making their choices regarding future drinking or use of marijuana more difficult.

With the current state of our knowledge, it may be an appropriate and justifiable policy to suggest to recovered heroin addicts who wish to drink socially that they may attempt to do so only after some arbitrary period of post-treatment stability, provided the manner of the use is reasonable and there is full awareness of the risks involved.[36,37]

Some drug programs at the present time allow residents to begin to drink socially, if they wish to, during the final or reentry phase of treatment. The rationale for this policy is that many, if not most, former heroin users are going to drink at some time, no matter what the program advised at the time of active treatment, and it is better for the experiments with alcohol use to take place while the person is still in treatment, with prompt help, if needed, at hand.

One can think of arguments for and against a policy of this kind, but we know of no hard evidence to decide the question objectively at this time. Where drug addicts and alcoholics are treated together, this kind of issue can be difficult and divisive.

At the top of the list of problems facing a combined treatment program as a whole is the "image problem" the program almost certainly will develop. Even though the problem is in the eye of the beholder, which is to say those outside who take a discouraging view of combined treatment, the unhappy consequences are felt by the program itself. At a federal level, the combined program is likely to encounter difficulty gaining recognition and support, since it does not fit within the categorical boundaries and definitions established by federal laws and maintained by categorically separate national institutes and their programs.

This problem has been lessening in the last few years; yet, a combined program is not at home in either the "alcohol camp" or the "drug camp." This is becoming less of a problem at the state level, since a majority of states, despite persistence of categorical separation nationally, now have administratively combined state alcohol and drug programs.

The problem is more serious closer to home, where the failure of the combined program to be seen as "appropriate" by "drug" or "alcohol" programs may deprive the program of support in the form of referral of clients. The alcohol constituency, for example, may withhold its approval and support, because many recovered alcoholics and many affiliated with AA perceive drug addicts as different and not to be treated in the same program with alcoholics. Years of documentation from actual experience and patient interactions with the leaders of these groups will be required before significant changes in attitudes occur.

IS COMBINED TREATMENT EFFECTIVE?

The question properly inquires whether combined treatment works and also how well it works compared to other available approaches. These questions are not easily answered, requiring as they would, carefully designed and executed comparative studies of combined and segregated approaches. Many problems would have to be overcome with respect to the selection of study samples, conducting the study in an acceptably "blind" or objective manner, and using appropriate criteria and procedures to assess positive treatment effects.

Since many variables may influence the outcome of treatment—in addition to combined or segregated treatment—their effects would have to be minimized through appropriate research design and/or statistical controls. This objective can, at best, be achieved in only a relative sense.

Some of the more important intervening variables that should receive special attention in any comparison of combined treatment and segregated treatment include: age, socioeconomic status and other personal characteristics; patterns and duration of substance abuse; presence of irreversible complications at the time of treatment; the mortality rate in all the groups under study; the experience and skill levels of the respective staffs providing treatment; the length of time the treatment program has been operating as a combined or segregated program; the composition of the respective staffs in terms of the number of professionals and paraprofessionals; the racial and sexual composition of the staffs; and the emotional and motivational status of the respective staffs delivering services.

Suffice it to say that one cannot arrive at valid conclusions by comparing the reported outcomes of different programs carrying out different treatments on different populations at different times unless adequate controls have been established to permit such comparisons. In a similar vein, one cannot even validly compare the outcome of treatment of alcoholics to the outcome of the same treatment of drug addicts, even though they are treated together in the same program and at the same time, unless one can first establish that the two groups were appropriately matched with respect to variables that could significantly influence treatment success.[38]

If one group, eg, the alcoholics, happened to represent the poor end of the scale of prognosis for all alcoholics, while the addict group was derived largely from the hopeful end of the addict distribution curve, no valid comparison could be made. Given the innumerable factors, known and unknown, that determine which persons come into any treatment programs, it is practically impossible to ensure true equivalence of different patient populations for purposes of comparison.

Despite all of the methodological difficulties cited, two groups of investigators have carried out studies designed to compare combined and separate treatment of alcoholic and drug addicted patients.[39,40] In neither study did the results strongly indicate superiority of outcome for either combined or segregated treatment. The Eagleville study[39] supported the conclusion that combined treatment has therapeutic effects that are no less than those of segregated treatment. The Veterans Administration Study[40] concluded that combined treatment had therapeutic effects, but slightly less than those for segregated treatment. We consider neither of these studies to be definitive; each had its problems related to the issues discussed earlier.

In view of the difficulties that must be overcome to conduct a valid comparison study of this kind, and the considerable cost both direct and intangible, a definitive study seems improbable. We are left, therefore, with the meager—and somewhat divergent—results of the studies already completed, plus what we can derive from the experience with these treatment approaches in nonexperimental conditions. We do not believe, moreover, that a detailed analysis or vigorous critique of the work reported to date would change this conclusion. In lieu of any further discussion, we refer readers to available reports.[34,39,40]

CONCLUSIONS

For the present, our opinion is that combined treatment is an effective approach when properly used, and that the results obtained with combined treatment compare favorably with those obtained with other approaches. As with any treatment, the skill of those who practice it makes a difference, and experience sharpens skill. And, as with other treatments, it works better for some people than for others. The selection of those most suitable for this approach has not been worked out with any precision. In this chapter we have indicated what kinds of persons have fared less well in our programs.

We see no reason to discontinue Eagleville's combined programs, believing that, for us, the advantages outnumber the disadvantages. We advocate that others try some combined approaches. For those hearty souls willing to accept our offer, we have stressed the importance of whole-hearted efforts and have discussed some critical issues concerning staff and their training.

We believe it is correct and very useful to view alcoholism and other substance abuses and addictions as part of the same problem. A generic concept provides a framework that can accommodate and encompass current epidemiologic reality and serves well for the study of today's substance-abuse problems and the planning of organized responses. For the professional in the field, generic conceptualization and training along generic lines offer the best opportunity to stay abreast of changing patterns of substance abuse and to meet the needs of persons requiring treatment.

REFERENCES

1. Carroll JFX, Klein MI, Santo Y: Comparison of the similarities and differences in the self-concepts of male alcoholics and addicts. *J Consult Clin Psychol* 46:575–576, 1978.

2. Cohen A, Barr HL, Ottenberg DJ: Comparing heroin addicts and alcoholics on the Personality Research Form. Unpublished manuscript, 1978. (Available from HL Barr, Eagleville Hospital and Rehabilitation Center Research Dept, Eagleville, PA 19408.)

3. Carroll JFX: Similarities and differences of personality and psychopathology between alcoholics and addicts. *Am J Drug Alcohol Abuse* (in press).

4. Carroll JFX, Malloy TE, Roscioli DL, Godard DR: Personality similarities and differences among four diagnostic groups of female substance abusers. *J Studies Alcohol* (in press).

5. The Committee on Professional Relations of the General Service Board of Alcoholics Anonymous. *About A.A.: A newsletter for professionals and semiprofessionals,* Summer 1972, p 2.

6. Pittman DJ: The rush to combine: Sociological dissimilarities of alcoholism and drug abuse. *Br J Addict* 64:337–343, 1967.

7. Neumann CP, Tamerin JS: The treatment of adult alcoholics and teenage drug addicts in one hospital: A comparison and critical appraisal of factors related to outcome. *Q J Stud Alcohol* 32:82–93, 1971.

8. Ottenberg DJ: The reluctance to combine. *Am J Drug Alcohol Abuse* 4:279–291, 1977.

9. Smart RG, Fejer D: Six years of cross-sectional surveys of student drug use in Toronto. *Bull Narcotics* 27:11–22, 1975.

10. Kandel D, Faust R: Sequence and stages in patterns of adolescent drug use. *Arch Gen Psychiatry* 32:923–932, 1975.

11. Hamburg BA, Kraemer HC, Jahnke W: A hierarchy of drug use in adolescence: Behavioral and attitudinal correlates of substantial drug use. *Am J Psychiatry* 132:1155–1163, 1975.

12. Bogg RA: Drinking as a precursor to hallucinogenic drug usage. *Drug Forum* 5:55–67, 1975–1976.

13. Gould LC, Berberian RM, Kasl SV, et al: Sequential patterns of multiple-drug use among high school students. *Arch Gen Psychiatry* 34:216–222, 1977.

14. Beschner GM, Friedman AS (eds): *Youth Drug Abuse.* Lexington, Mass, Lexington Books, 1979.

15. Carroll JFX, Malloy TE, Hannigan PC, et al: The meaning and evaluation of the term multiple substance abuse. *Contemp Drug Problems* 6:101–134, 1977.

16. Gerston A, Cohen MJ, Stimmel B: Alcoholism, heroin dependency, and methadone maintenance: Alternatives and aids to conventional methods of therapy. *Am J Drug Alcohol Abuse* 4:517–531, 1977.

17. Kaufman E: Polydrug abuse or multidrug misuse: It's here to stay. *Br J Addict* 72:339–347, 1977.

18. Carroll JFX (ed): *National Drug/Alcohol Collaborative Project (NDACP) Final Report* (NIDA Grant H81 DA 01113). Rockville, Md, National Institute on Drug Abuse, 1977.

19. Carroll JFX: Uncovering drug abuse by alcoholics and alcohol abuse by addicts, in Levin M (ed): *Proceedings ADPA 28th Annual Meeting.* Washington, DC, The Alcohol and Drug Problems Association of North America, 1977.

20. Barr HL, Cohen A: *The problem-drinking drug addict.* (DHEW Publication No. [ADM] 79-893). National Institute on Drug Abuse, Services Research Report, US Dept HEW, Public Health Service, ADAMHA, 1979.

21. Pascarelli EF, Eaton C: Disulfiram (antabuse) in the treatment of methadone maintenance alcoholics, in DuPont RL, Freeman RS (eds): *Proceedings of the Fifth National Conference on Methadone Treatment.* New York, NAPAN, 1973.

22. Williams WV, Lee J: Methadone maintenance: A comparison of methadone treatment subjects and methadone treatment dropouts. *Int J Addict* 10:599–608, 1975.

23. Dole VP, Nyswander ME: Methadone maintenance treatment: A ten year perspective. *JAMA* 235:2117–2119, 1976.

24. Bihari B: Alcoholism and methadone maintenance. *Am J Drug Alcohol Abuse* 1:79–87, 1974.

25. Maddux JF, Elliott B: Problem drinkers among patients on methadone. *Am J Drug Alcohol Abuse* 2:245–254, 1975.

26. Gearing FR, D'Amico DA, Tousie S: Does Methadone Maintenance Lead to Increased Alcohol and Polydrug Use? Paper presented at the Third National Drug Abuse Conference, New York, March 1976.

27. Zampanis T (panelist): Total abstinence—Is it the only way? in Carroll JFX, How Eagleville came to this conference topic. *Am J Drug Alcohol Abuse* 5:257–275, 1978.

28. Ottenberg DJ, Rosen A: Merging the treatment of drug addicts into an existing program for alcoholics. *Q Stud Alcohol* 32:94–103, 1971.

29. Driscoll GZ: Comparative Study of Drug Dependent and Alcoholic Men at Eagleville Hospital and Rehabilitation Center. Paper presented at the

Alcohol Drug Problems Association Annual Conference, Hartford, Conn, September 1971.

30. Driscoll GZ, Barr HL: Comparative study of drug dependent and alcoholic women, in Brock R (ed): *Selected Papers Presented at the General Sessions Twenty-Third Annual Meeting.* Washington, DC, The Alcohol and Drug Problems Association of North America, 1972.

31. Ottenberg DJ: The Evolution of Eagleville's Combined Treatment Program. Paper presented at the National Drug Abuse Conference, Chicago, March 1974.

32. Carroll JFX: "Mental illness" and "disease"—Outmoded concepts in alcohol and drug rehabilitation. *Commun Ment Health J* 11:418–429, 1975.

33. Carroll JFX: Mental illness and addiction: Perspectives which overemphasize differences and undervalue commonalities. *Contemp Drug Problems* 7:227–231, 1978.

34. Ottenberg DJ: Combined treatment of alcoholics and drug addicts: A progress report from Eagleville. *Contemp Drug Problems* 4:1–21, 1975.

35. Carroll JFX: Staff burnout as a form of ecological dysfunction. *Contemp Drug Problems* 8:207–225, 1979.

36. Carroll JFX: How Eagleville came to this conference topic, in Ottenberg DJ, Carpey EL (eds): *Proceedings of the 10th Annual Eagleville Conference: Abstinence and Controlled Use as Treatment Objectives for Alcoholics and Substance Abusers. Am J Drug Alcohol Abuse* (special issue)5:257–275, 1978.

37. Carroll JFX: Does sobriety and self-fulfillment always necessitate total and permanent abstinence? *Br J Addict* 75:55–63, 1980.

38. Barr HL, Rosen A, Antes DE, et al: Two year follow-up study of 724 drug and alcohol addicts treated together in an abstinence therapeutic community, in *1973 Proceedings: Fifth National Conference on Methadone Treatment.* New York, National Association for the Prevention of Addiction to Narcotics, 1973.

39. Aumack L: Evaluation of Eagleville's residential combined treatment program, in Carroll JFX (ed): *National Drug/Alcohol Collaborative Project (NDACP) Final Report* (NIDA Grant H81 DA 01113). Rockville, Md, National Institute on Drug Abuse, ch 15, pp 15-1–15-98, 1977.

40. Veterans Administration. Evaluation of the Pilot Alcohol and Drug Treatment (PADAT) Project. Unpublished manuscript, 1977. (Available from Veterans Administation, Department of Medicine and Surgery, Washington, DC 20420.)

7 Prevention of Alcoholism and Drug Abuse

John D. Swisher
Judith R. Vicary

At the First Annual Alcohol, Drug Abuse, and Mental Health Administration Conference on Prevention sponsored by the federal government in 1979, Dr Helen Nowlis from the United States Office of Education opened her remarks by stating:

> The concern of this session is commonalities and differences in alcohol *and* drug abuse *and* mental health prevention programs for children and youth. Ten years ago a session on this topic, indeed this conference itself, would have been highly improbable, if not impossible.[1]

Nowlis's view of the probabilities of considering the similarities and differences in alcohol and drug abuse prevention is still accurate today. The recommendations from that conference called for further exploration of the commonalities, with the hope of stimulating interagency sponsorship of comprehensive projects. However, very few, if any, such projects have come to fruition, and the agency boundaries which focus on differences have persisted.

On a community level, however, there has been a resurgence of concern for prevention of both drug and alcohol abuse on a coordinated basis. The concern has taken many forms, from enforcement to ad hoc advisory groups, within both schools and other community settings. While school-age populations are still the primary target audience, adult populations are beginning to be of concern. Business and industry employees, women and the elderly are targeted in the new programs that have been initiated. New efforts are also being made with adults who work with youth. Several states, for example, are helping parent groups to organize and develop prevention strategies.

Surveys of alcohol use and abuse among youth reveal a continuing high level of consumption for the past several decades.[2] The Drug Abuse Warning Network, based on data from drug abuse crises reported in emergency rooms, also notes high levels of abuse for a number of substances, particularly in combination with alcohol.[3] A continuing need for prevention activities appears justified.

In the mid-1960s, when there was an earlier public focus on alcohol and drug abuse, the first responses to the problems emphasized the provision of information about the (dangerous) effects of drugs and alcohol to young people. The assumption was made that if students were educated about drugs and alcohol they would choose not to abuse these substances. Ex-addicts and alcoholics, law enforcement agents, pharmacists, and physicians were called on to educate masses of students, often with fear approaches.

Evaluations of this type of program concluded that an emphasis on information was an insufficient strategy and/or harmful, in that some of these programs increased levels of use among youth.[4] Drug educators, more than alcohol educators, began to emphasize approaches that attempted to foster self-understanding, self-esteem, and interpersonal relations among youth. These strategies have been broadly labeled affective or humanistic approaches. Concurrently, it appears that many drug abuse prevention programs either completely abandoned, or at best, greatly reduced any coverage of the cognitive aspects of drug use/abuse. Conversely, alcohol abuse prevention programs continued to focus on imparting information, and either excluded affective strategies or used them only in the context of decisions about drinking.

PRIMARY PREVENTION: ACTIVITIES AND GOALS

Drug and alcohol abuse prevention are those planned-for activities that occur *before* the onset of misuse, immediate crises, or sustained problems of abuse with various substances. The appropriate audiences for this type of program include all ages because throughout the life span there are developmental stages that require new or recurring decisions about substance-use behavior. It should be emphasized, however, that most concern and, therefore, programming, has been directed at youth populations. A continuing problem in the field of primary prevention is the lack of clearly defined goals, particularly in terms of desired impact on an intended audience. Braucht et al[5] noted that, despite the almost universal state requirements for schools to include alcohol and drug education in the curriculum, most published curricula at that time failed to articulate goals.

The meaning of the concept "abuse" has also been unclear because of the various professionals that are involved in the field.[6] For example, medicine has defined abuse as occurring when an individual fails to follow a prescription or engages in self-medication. Law enforcement defines abuse as any use of illicit drugs, whereas the social sciences define abuse as use that is harmful to the individual or to society. For purposes of this chapter, the social sciences position is the context in which the term abuse is defined.

An implicit goal of many early programs was total abstinence from illicit drugs and/or alcohol. However, given the natural curiosity of youth and the availability of many substances, as well as the acceptance by many adults of alcohol and licit drugs, total abstinence as a goal was not possible.

The objective of teaching for responsible individual choice regarding use of various substances has recently emerged as a reasonable and achievable outcome for prevention programs. This position defines responsible choice as use that results in very few negative consequences for the individual and society, such as crises, injury, addiction, occupational dysfunction, related diseases, etc.

Blum,[7] based on a three-year study, recommended:

> It [the public school] can reasonably target highly disapproved particular compounds as requiring rejection (eg, amphetamines, heroin, inhalants). The school can also encourage controlled rather than extreme use of given substances such as alcohol by communicating knowledge of the conditions under which moderate use is learned and unsafe outcomes are avoided.

Table 7-1 compares alcohol and drug abuse prevention activities in terms of the extent of use and the general category of social response. In any primary prevention effort it is important to make some distinctions among the various types of activities and the categories of responses. Such distinctions will subsequently allow for the rational development of primary prevention objectives separate from equally important, but potentially confusing, activities (eg, law enforcement).

While the table represents the separate categories of activities, overlapping relationships, such as the deterrent effects of law enforcement as a form of primary prevention, are also recognized.

Table 7-1
Comparison of Alcohol and Drug Abuse Prevention Activities

		Examples of	
		Drug Abuse Activities	*Alcohol Abuse Activities*
Primary	Abstinence or experimentation	Information Alternatives Media campaigns	Alcohol-specific printed material Traffic safety Media campaigns
Secondary	Excessive and/or episodic use	Crisis intervention Early diagnosis Referral	Emergency care Occupational programs Referral
Tertiary	Chronic abuse	Methadone maintenance Detoxification Therapeutic community	Disulfiram Detoxification AA
Judicial System*	Intoxication Sale or distribution	Undercover investigations Expungement of records Adjudication to treatment facility	Breath analyzer DWI programs

*Law enforcement and court activities overlap with primary, secondary, and tertiary activities.

If responsible individual choice were accepted as an objective, it would necessitate more balanced prevention efforts than presently exist. There would be a need to provide individuals with accurate information and to assist them with their personal development in order to prevent the use of alcohol or drugs from becoming a prolonged expression of some more basic developmental problems.

THE EFFICACY OF INFORMATION APPROACHES

The role of information about drugs or alcohol in affecting an individual's decision to use/abuse drugs or alcohol remains highly controversial. The early beginnings in prevention were guided by the philosophy: "We are there just to present the kind of information that is available so that they (the students) can figure out for themselves."[8] This type of simplistic thinking was subsequently challenged by a series of studies that questioned the emphasis on increasing knowledge.[4] The assumption made by an approach to prevention that emphasizes information would be:

1. Increased knowledge about drugs or alcohol;
2. Leads to conservative attitudes towards use;
3. Which in turn reduces use or prevents abuse of drugs or alcohol.

However, the opposite assumption may be equally valid:

1. Use of alcohol or drugs;
2. Leads to attitudes favoring use which in turn;
3. Guide the selective reception and retention of information.

Very few, if any, studies or programs have adequately addressed these assumptions. Some recent findings suggest either a negative impact or no impact from information programs.[9] Goodstadt et al concluded:

> Reported exposure to drug education was more common among those grades 7 and 9 students who reported more drug use—alcohol, tobacco or marijuana. This relationship was less strong for grade 9 than for grade 7 and was non-existent for grades 11 and 13....
>
> The relationship between exposure to drug education and frequency of reported drug use appeared to hold up for tobacco education and tobacco use among grade 7 students, and marijuana education and marijuana use among grades 7, 9 and 11; little evidence was found for any effect (either positive or negative) of alcohol education on frequency of alcohol use or on inappropriate alcohol use.[10]

Another problem complicating informational strategies for alcohol or drug abuse prevention has been the lack of credible sources of information. Reviews of films[11,12] concluded that three out of four films presenting facts about drugs or alcohol were scientifically inaccurate. Teachers using these materials obviously lost some credibility with certain knowledgeable members of their audiences. Others have found similar problems with printed materials,[13] and by the mid-1970s a federal moratorium was placed on the production of this type of material. This problem and others with media materials remain to such an extent that the National Institute on Drug Abuse postponed its national media campaign for 1980.

Finally, levels of information (eg, accumulation of facts *vs* synthesis of principles) also need to be studied. Higher level mental processes may indeed alter the impact of the knowledge on the intended audience. The use of a biology lab and the scientific method for studying biology have long been established as a better means for understanding biology as well as for creating more positive attitudes toward the use of science in everyday life. However, information oriented approaches in drug and alcohol education have not been adequately modified and/or evaluated in this manner.

THE RELATIONSHIP OF AFFECTIVE STRATEGIES TO PREVENTION

At the same time that the early results regarding the ineffectiveness of information in drug and alcohol abuse prevention became known, a nationwide movement was underway which centered around affective/humanistic strategies, eg, values sharing, decision making. These approaches were particularly appealing to school teachers in that the techniques frequently appeared to motivate students and revitalize the classroom. Many drug and alcohol abuse prevention personnel were exposed to these modalities through various in-service social problem or education workshops. However, the new methods were often adopted "as is," with no effort made to relate them directly to prevention goals or substance-abuse objectives.

Advocates of affective/humanistic techniques argued that students who experienced these programs would be less likely to experience developmental problems and, therefore, would also have fewer problems with substance abuse. Despite the enthusiasm expressed by the prevention field, and the often worthwhile efforts, questions were raised repeatedly by the drug and alcohol authorities at all levels as to what was appropriate to fund as a prevention program. These concerns included:

1. How effective are affective approaches for the prevention, specifically, of drug and alcohol abuse?
2. To what extent should one funding source be responsible for an approach that may be equally relevant to the prevention of other social problems, eg, delinquency?
3. Where do effective approaches as "prevention" merge with the ongoing programs of schools and other community agencies?

For the most part these questions have gone unanswered. For example, relatively little evaluation has been done regarding the outcomes and relationship of a purely affective approach to prevention. In general, the substance-abuse field has not:

1. Developed clear objectives and measurable outcomes for any affective element, eg, improving self-esteem.
2. Connected techniques to the accomplishment of objectives.
3. Related the affective variables to drug and alcohol abuse prevention objectives.

Many more subtle problems have also characterized the programs that emerged from this orientation. For example, the programs have not decided how much internal or external locus of control is helpful to their clients, or how durable are certain affective improvements, such as interpersonal communciations, over time. Both the National Institute on Alcohol Abuse and Alcoholism (NIAAA) and the National Institute on Drug Abuse (NIDA) are encouraging clearer conceptualization by programs of the relationship between affective development and substance abuse prevention. Furthermore, programs are being funded by either institute only if they incorporate extensive planning for evaluation.

Affective Overemphasis in Drug Abuse Prevention

One of the consequences resulting in part from the confusion about the place of information in drug abuse prevention was a shift away from information approaches toward those that included such techniques as values clarification, peer influence, alternative activities, self-understanding and self-acceptance, and interpersonal communication skills, among others. These techniques can be loosely categorized as methods used under the rubric of affective, humanistic, or psychological education. The programmatic directional shift resulted in the development of the following definitions at the federal level:

> Primary drug abuse prevention is a constructive process designed to *promote* personal and social growth of the individual toward *full human potential;* and thereby, inhibit or reduce physical, mental, emotional, or social impairment which results in or from the abuse of chemical substances.[14]

Similar definitions also emerged at the state level:

> Primary means that level of activity which is aimed at the total population. Such programs promote functional, positive, intra- and

> interpersonal skills, attitudes, and behaviors impacting upon the formulation of life styles which are likely to exclude substance dependency.[15]

The important point about these definitions is that they set the stage for drug abuse prevention programs that ignore drug-specific cognitive factors and emphasize the developmental affective factors.

Several states (eg, Georgia[16] and North Carolina[17]) developed prevention programs, under the sponsorship of agencies directly concerned with drug abuse, which make practically no direct reference to drugs. The title of the Georgia program is "Life Skills for Mental Health," and it assumes that the enhancement of "life skills" will ultimately result in fewer problems with drug abuse. The major goals are:

1. To help young people become more accepting of themselves and the significant people with whom they come in contact.
2. To help young people be more accepting of all feelings.
3. To help young people form more satisfying personal relationships with significant others.

Much to the credit of the program designers there has been a careful plan for evaluation, and the impact of this program as a preventive modality in drug abuse prevention will be assessed. The Georgia program is one of the most comprehensive and thorough in its development and in terms of evaluation, but it does not plan for "life skills" learning to be transferred by the individual to drug-specific situations, nor does it impart drug information. The program will probably *test* for self-initiated transfer of learning, but if the results are inconclusive, it will not have allowed for an analysis of the interaction between increased affective functioning (life skills) and the applications of these learnings to specific drug abuse situations young people will encounter.

The Georgia statewide program is characteristic of other statewide programs as well as of many local community or school-based efforts. While it is atypical in terms of total coordination, conceptualization, and evaluation, it is representative of the trend among drug abuse prevention projects to overemphasize the affective domain and deemphasize the cognitive domain relevant to substance information.

Rather than developing their own materials, many prevention programs also adopted existing affective education curricula and kits, using them without adding any reference to drugs. For example, Guidance Associates publishes several kits titled *Developing Understanding of Self and Others,* which are designed to improve self-esteem and interpersonal relations at the elementary school age level. These materials were typically used in prevention efforts without relating them specifically to drug use. Prevention, through enhancing positive affective development in order to prevent negative substance abuse behavior, was the goal. While many of the affective education materials are well planned and developed in their own right, their wholesale use as *the* prevention modality is probably not comprehensive enough to guarantee drug abuse prevention objectives.

COGNITIVE OVEREMPHASIS IN ALCOHOL ABUSE PREVENTION

From a historical perspective it is probably accurate to suggest that alcohol abuse prevention programs have emphasized the cognitive (informational) aspects of alcohol as the dominant component of most prevention programs. However, as Mong states, "Also, the extent of recent new knowledge about alcohol and alcohol abuse prevention has rendered obsolete most curricula developed before 1973."[18] At the present time there is a clear interest in and trend toward the inclusion of affective strategies in alcohol abuse prevention, but most alcohol prevention programs still continue to overemphasize the cognitive. The current "official" definition regarding alcohol abuse prevention, preferred by the National Center for Alcohol Education, reads:

> The purpose of prevention is to increase the likelihood that individuals will develop drinking-related behaviors that are personally and socially constructive. Negatively stated, prevention programs are aimed at reducing the number of persons whose alcohol-related behavior affects the way they carry out the roles and responsibilities of everyday living.[19]

In its own right this definition focuses on drinking behavior, but, by contrast to the drug abuse prevention definitions, this concept is very alcohol-specific. Furthermore, the means recommended to accomplish this objective more clearly emphasize the cognitive domain. For example, CASPAR, one of three models identified by NIAAA lists the following objectives:

- Increase student's factual knowledge about alcohol.
- Engender the development of mature, responsible feelings and values concerning alcohol use and abuse.
- Achieve long-lasting behavioral effects, resulting in the student's use or non-use of alcohol as an integrated and

positive feature of his/her living, free of personal or social problems.[19]

Obviously, the first goal is the task of increasing information about alcohol. Furthermore, the teacher training necessary to implement this model lists the following topics:

- Definitions of drinking, drunkenness, and alcoholism.
- Role of alcohol in American society.
- What alcohol is; facts and fiction about drinking.
- Physiological and psychological effects of drinking.
- Goals in alcohol education.
- Methods of teaching individual decision making about drinking.
- How students use alcohol; current research on youthful drinking practices.
- Alcoholism as a family illness and a public health problem.
- Nature of local community resources.[20]

A further example is of content analysis on a school-grade basis (Table 7-2) which reveals the dominant role of information throughout one of NIAAA's model programs.[21] The two affective components in this analysis are shown by some coverage of self-concept in the fourth grade, and coping skills at the fifth grade and junior high levels.

Table 7-2
Content Analysis of a Model Alcohol Abuse Prevention Program

	Grade Levels				
Skills	*4*	*5*	*6*	*Jr. High*	*Sr. High*
Knowledge	X	X	X	X	X
Self-concept	X				
Coping skills		X		X	
Attitudes			X		X
Self-reported drinking behavior				X	X

BALANCED AFFECTIVE AND COGNITIVE APPROACHES TO PREVENTION

Making a responsible choice about the use of alcohol or drugs involves a sense of responsibility for one's self, for others, and for society. Contemporary alcohol and drug abuse prevention programs have attempted, albeit piecemeal, to focus on these aspects of responsibility.

The elements that are typically included in affective techniques are:

1. Increasing one's self-awareness, self-esteem, and responsibility for one's acts.
2. Increasing one's interpersonal skills and responsibility for interaction with others.
3. Increasing one's ability to cope with and responsibly participate in society's institutions.

These three dimensions of responsibility in affective development form a core of skills which, if fully developed, supposedly will enhance an individual's total development and will, therefore, reduce interest in, need for, and severity of reactions to a host of substances. Enhanced development, according to the most optimistic, will also lead to a reduction in other social problems as well (eg, delinquency).

Carney[22] and Blum[7] concluded that information programs *combined* with personal and interpersonal growth were more likely to be successful. Apparently it is important to examine one's values, decision-making skills, susceptibility to peer pressure, etc, in the context of information about drugs and alcohol. More specifically, it might be possible to teach the steps in effective decision making, but unless those steps are also applied to decisions an individual might face regarding substances, the decision-making skills will not transfer to the context of substance abuse. Regardless of the personal and interpersonal skills, it is equally important to assist individuals with examining the implications each affective component has for drug and alcohol abuse prevention.

This very brief definition of an affective approach is unfortunately still ambiguous. This is particularly true when it is recognized that little antecedent effort has been directed at clarifying this concept or any of the elements within it. In fact, some prevention professionals would resist further delineation of the concepts for fear of losing freedom or stifling their creativity. From an administative perspective, however, it becomes very difficult to say what is and what is not an appropriate prevention activity. The question of which agency's responsibilities these are, and the subsequent funding issues, can result from this lack of clarity.

Despite these ambiguities, enriched personal and social growth probably are essential to any prevention effort that advocates responsible choosing. Enhanced development as part of a total primary prevention effort is recommended on the basis of basic findings in program evaluation. First, straight information about a social problem (eg, alcohol or smoking) has not been demonstrated to be effective in preventing a problem, and some researchers believe these information dissemination approaches can even be potentially harmful. Second, recent correlational research[23] has demonstrated highly significant relationships between tests of self-esteem, understanding of others, social responsibility, and coping with change; and abusing use of marijuana and alcohol.

Third, there are some programs in which personal development was improved and drug and alcohol attitudes and/or behavior were simultaneously improved. As examples:

- In a large city system when the students were assisted with decision making, problem solving, and alternatives, their alcohol and drug use decreased and school achievement increased.[24]
- In a small town junior high school, when the students' decision making was improved, their use of alcohol and drugs was reduced.[25]
- In a city school system, when the personal development of fifth and seventh grade students was maximized and their knowledge of alcohol and drugs was increased, there was a reduction in their use.[22]
- In a group of college students, when they were shown discrepancies between desired social activities and drug or alcohol preferences, their attitudes toward use became more conservative.[26]
- In a small town, when parents of young children were taught more effective ways of relating to their children, the parents felt they could aid in the prevention of drug and alcohol problems.[27]

The scientific adequacy of these studies is far from perfect, but they consistently found that enriched personal and/or social development coupled with accurate information may be effective abuse prevention efforts. Reviews of research in prevention indicate little success,[28] but the general conclusion is that programs that combine affective development with accurate information about substances are the most promising and should be encouraged.

Most of the programming and subsequent evaluations have only dealt with school-age youth. They have attempted to reach young people through a variety of significant others (peers, parents, and teachers), but the ultimate target audience has been the youth population. The remainder of this chapter discusses the implications of a balanced model of prevention for business and industry and for the elderly.

NEW PREVENTION AUDIENCES

Occupational Alcoholism Programs

While the previously described prevention efforts have focused exclusively on youth, it has been increasingly recognized that drug abuse, as well as alcohol abuse, are not limited by age, and serious problems exist also with adults. Some of these misuse/abuse behaviors are a continuation or exacerbation of problems begun during school years. But an even larger number of adults who began their misuse as adults, even as late as in their fifties, sixties, or seventies, are being identified at intervention and/or treatment stages. Consequently, in recent years prevention efforts have been suggested as necessary for a variety of adult populations, to be designed for and located in settings appropriate to their age and activities.

The occupational setting is the most logical environment in which to reach large numbers of adults, presumably prior to dysfunctional abuse behaviors developing. The historical perspective is also instrumental in this approach since many corporations already have well-established occupational alcoholism programs which operate on secondary and/or tertiary levels. Major efforts at developing industrial alcoholism programs began in the 1940s and 1950s, pointing out the costs to business and industry, as well as to society, of the abuse of alcohol. Increased absenteeism, poor judgment, reduced efficiency, lost time, and morale problems were among the factors cited that cause company losses annually of billions of dollars. Another major fact used to persuade company officials to develop programs and procedures was the investment they had in trained, experienced employees who would be an expensive loss in their middle service years. The National Council on Alcoholism (NCA), a voluntary organization begun in 1944, and the Yale Center of Alcohol Studies (now at Rutgers University), founded in 1941, pioneered efforts to persuade several companies, eg, DuPont, Allis Chalmers, and Consolidated Edison, to establish alcoholic worker identification and treatment referral programs.

Most alcoholism programs originating then were narrowly based projects which barely dealt with the full range of ailments and disorders that accompany alcoholism. Alcoholics Anonymous (AA) programs were usually the only sources for referral, counseling, and treatment. Another limitation was that management and unions did not work together in these initial alcoholism programming attempts. In fact, it appears that each of these two groups overlooked or neglected the important role of the other in unionized companies, and consequently reduced potential programming impact. The medical departments of participating companies usually had the responsibility for the implementation of the early programs.

Dr Ralph Henderson, a pioneer in the development of alcoholism programming in industry, was one of the first to suggest that threatened job loss due to poor job performance resulting from alcohol problems would be a strong motivator for an employee with a drinking problem to seek help, and to continue with a rehabilitation program. He was a major force in the efforts to convince employers and management level personnel to begin programs within their companies.

In the 1950s, research began to show the full impact of alcoholism on job impairment. At that time, dependencies of other sorts, and emotional disturbances in general, were not addressed in company policies. The use of "impaired job performance" as a rationale for employer intervention began to grow and as it did, in the 1960s, programs became less narrow and widened their focus of attention to more generic problems.

The NCA, in the 1960s, expanded its efforts with employers to develop new occupational alcoholism projects, and the number of programs increased more than 600% during that decade. Another significant factor in the growth of occupational efforts to combat alcoholism was the passage of the Hughes Act, which established in 1970 the National Institute of Alcoholism and Alcohol Abuse (NIAAA), including an Occupational Programs Branch. NIAAA also helped widen the programmatic emphasis by endorsing the "broad-brush" or employee assistance concept, in which any employee with poor work performance related to health and/or emotional problems, not just alcoholism, was identified and referred for treatment. With job performance as the basis for intervention, strong incentive is provided for accepting the helping services offered; continued employment depends on participation and improved job behavior. This philosophy is seen as particularly valuable in intervening with workers before their drinking problems become so severe that rehabilitation efforts are unlikely to be of much use. The traditional "five r's" have been the basic steps most programs employ: 1) recognition, 2) respect, 3) referral, 4) restoration, and 5) readjustment.

Another organization currently instrumental in alcohol abuse occupational program development is the Association of Labor-Management Administrators and Consultants on Alcoholism (ALMACA). From its beginning in 1971, this organization has grown from 12 to over 1100 members, representing service providers in various industrial alcoholism programming. Its growth also parallels the recent increase in union- and business-sponsored employee assistance programs. ALMACA states:

> Whatever the size of your (business), you can be sure that it is losing money because of alcohol problems among your employees.
>
> We say (this) flatly on the basis of extensive studies that found the American economy losing more than $25 billion each year...in lost production, waste, high medical costs, absenteeism...all attributable to alcohol-related problems.
>
> Your organization shares in that annual loss.[29]

The success of company union-sponsored, work-based programs varies considerably, from a 50% to 70% rate, with 66% cited as the average industry success rate nationwide.[30] Twenty-two non-work-setting evaluations showed the majority of programs averaging from 18% to 35%.[31] Trice and Roman[32] noted that if rehabilitation other than job retention alone is the success criterion, company programs show rates of 50% *vs* 20% for state hospital programs. Earlier identification, intervention, and treatment, as well as continuing employment and job stability, are the factors credited with these higher success statistics. Confrontation on poor job performance also appears to be a powerful motivator.

Occupational Drug Programs

During the late 1960s, when drug abuse problems increased at great rates in the United States, the "drug scene" also invaded the employment field. Industry's position toward drug abuse among its employees initially was denial. Urban[33] found that retaliatory, punitive measures followed the denial phase and later were usually replaced by either in-house or extra-company counseling and, occasionally, follow-up procedures. These were usually an expansion of existing alcoholism programs and were renamed "alcoholism and other substance abuse" projects. This concern for drug-related employment problems caused both management and labor groups to reexamine their current employee alcoholism programs, the results, and their possible applicability as total substance-abuse procedures.

In 1970 several large companies sponsored the "First Symposium on Drug Abuse in Industry" and addressed issues such as policy formation and descriptions of sample policies for various size companies. They also examined drug screening procedures and treatment options. Efforts in this area initially were limited to large East Coast metropolitan areas; Urban[33] found that businessmen in other areas either refused to acknowledge the increased exposure to drugs of the people within their employment or felt that they did not have a substance-abuse problem. At approximately the same time, unions were beginning to address drug abuse among their members. Also in 1970, the AFL Community Services Committee sponsored one of the first union-based drug seminars. Drug information campaigns remained the focus of this committee for the next several years.

In an attempt to determine the extent of drug abuse in the labor force, the Research Institute of America surveyed 80 New York area companies in 1970.[34] Of these, 90% reported incidences of drug abuse within the company which had resulted in thefts, higher insurance rates, poor work performance, and increased absenteeism. There was also an expectation expressed that they would find many more abusers in their work forces within the next few years. Another study, reported in 1971 by the New York State Narcotic Addiction Control Commission,[35] examined drug use in that state's labor force. Significant rates of regular drug use were found in all occupational groups except farmers. Sales workers, for example, reported 12.3% used barbituates and 8.6% regularly smoked marijuana. Rush and Brown[36] studied 222 firms, 91 nonmanufacturing and 131 manufacturing, of which 53% had found drug abuse of some degree among their employees. Most also reported limited experience in dealing with the problem and over half expected the problems shortly to become more extensive.

Since 1971 the number of occupational programs that address substance abuse has grown from 300 to approximately 1500. While this rate of increase is commendable, it should be noted that there are over 1½ million businesses in this country and, therefore, the percentage with programs is still very small. In addition, many of these are not being adequately implemented. In 1974 it was estimated by Dr Steven Levy, director of Research and Program Planning of the Training for Living Institute in New York City,[37] that only 100 companies across the nation had active drug abuse programs.

The United States military forces represent another major employer that has recognized the cost of alcoholism and drug abuse among its personnel. It was estimated, for example, that in 1975 the cost of lost production in the military because of alcoholism and drug abuse was over $411 million.[38] The Department of Defense has begun a major program of education, rehabilitation, and treatment to combat both the lost productivity of its personnel and the high social cost. The military, like the private sector, generally emphasizes recognition of problems, intervention, referral, treatment, and rehabilitation. Inductees to the services are introduced to military policy on alcohol and drug use, identification methods, familiarization with health services available, training of supervisors to confront drug and alcohol situations, and long-term employment possibilities following successful treatment.

The extensive development of both corporate and military programs and policies dealing with the troubled employee, particularly in the areas of drug and alcohol abuse, has been primarily limited to intervention and treatment options. While some programs advertise the inclusion of education and/or prevention as part of their program, the range of these activities is extremely limited and would be better defined as awareness programs in light of current federal definitions of primary prevention. An example of this type of program theme can be seen in the Alcohol Awareness Education Seminar conducted by the United States Air Force.[39] Their eight-hour program is designed to give information on the entire range of alcohol and its related use, promote a self-awareness of individual drinking habits, and emphasize a concept of responsible drinking. It was developed in 1975 by the Air Force's Department of Social Actions Training at Lackland Air Force Base in Texas. It parallels early school drug education curricula in its emphasis on pharmacological and legal aspects of alcohol use. Information is given about the scope and impact of alcohol use in the United States, definitions of problem drinkers and alcoholics, effects of alcohol, stages of alcoholism, and ways to use alcohol wisely. However, one of the ten sessions also does deal with the affective dimension through values clarification, emphasizing the development of personal values and knowledge about one's own behavior in relation to alcohol.

Prevention efforts in industry include employee mailings, posters in the work environment, and other informational material that gives factual data about the dangers of drug and alcohol misuse. Another type of education program is one in which managers and employees are shown films and hear speakers in order to make them more aware of the various types of drug-related behaviors among employees. These awareness programs are often a first step in involving the participants in the intervention and referral processes. An example of this type, reported by Payne et al,[40] was conducted by the Spokane Regional Drug Abuse Training Center. The 60 staff participants, selected by management from a large northwestern industrial corporation, received nine hours of training. The impetus for the workshop was on-the-job personal safety hazards due to drug abuse. "Prevention" subjects included identification, pharmacology, behavioral effects, crisis intervention, and workshop planning.

In summary, it appears that when the term substance-abuse prevention is used in business, industry, and labor programs, it usually means the prevention of *greater* problems, both personal and job related, for the particular employee. However, the concept of preventing specific problems *before* they occur, or of promoting positive life skills, does not seem to be addressed in most employee assistance programs to this point.

Unique Occupational Prevention Programs

However, several new and innovative programming attempts in occupational areas have been begun within the past few years. A study of health programming in work settings on topics other than substances has shown that a more focused primary prevention approach is beginning to be used; fitness or exercise programming, nutrition projects, stress workshops, and weight reduction clinics are examples seen in a wide variety of companies.

In addition, the significance of health promotion or total prevention programs is now being recognized by both management and union officials. In addition to a humanitarian rationale for "wellness," prevention is being examined relative to cost effectiveness, an important consideration in business and industry programming. While these efforts are not designated as drug and alcohol abuse prevention programs, their more generic approach incorporates some of these same primary prevention philosophic and programmatic directions, as well as identifies some substance-specific issues. A brief overview of representative approaches to health promotion by business and industry will identify the several current, and promising, generalized approaches to substance-abuse prevention for adult populations.

Industry in America is now taking a major lead in the development and financing of health promotion, primarily to reduce the escalating costs of employee illness and health care. Many corporations, for example, are offering company-based Health Maintenance Organizations (HMOs) for their employees, or are encouraging their participation in a community-based HMO. The HMO one-fixed-fee approach provides total health care from preventive to hospitalization, and encourages, therefore, earlier and more complete preventive care.

Another direction seen in corporation settings is the preventive nature of a variety of medical clinics and/or physical fitness programs offered. As long ago as 1894 an employee fitness program was developed in a corporate setting at the National Cash Register Company in Dayton, Ohio.[41] There, President John Paterson began morning and afternoon exercise breaks for the workers and later added a company gym and park. However, only recently has there been a real interest among American business executives in such efforts.

According to recent figures[42] more than 400 major corporations as well as numerous small ones have exercise programs for some or all of their employees. Xerox Corporation, for example, has eight fitness centers, run by a full professional staff. Boeing Company has been developing fitness programs at its six Seattle area plants since 1970. Calisthenics, including flexibility and stretching exercises, are combined there with an aerobic running program, involving more than 2000 total program participants. While only limited data are available, preliminary studies indicate that such programs are beneficial in corporate terms: at Northern Natural Gas Company in Omaha, it was found that significantly fewer days were lost to sickness for those participating in the aerobic program than for those not involved; and at Occidental Life Insurance Company of California regular company gym users were absent only half as often as nonparticipants.[43]

More complete health programming on a prevention level has also been developed in many major companies. One of the most complete corporate health models, frequently cited as a model program, is run by Kimberly-Clark Corporation in Neenah, Wisconsin.[44] Their approach to health care emphasizes wellness, not illness, and is paid for entirely by the firm. Health maintenance, or improvement, rather than medical assistance after illness occurs, is the primary goal. Effectiveness is being measured through an extensive computerized health history used to document changes in health status. The insurance carrier is also working with the corporation to compare the hospitalization costs and incidences of major illness for program participants with the same data for a control group not included in the program. Drug and alcohol use information are among the subjects included.

Another model program has been developed by the Kansas Department of Health and Environment: PLUS is a Program to Lower Utilization of Services, a low-cost employee health improvement program.[45] It was designed for business and industry to offer as a benefit to their employees, in order to keep workers well and performing productively on the job. This health education program emphasizes for each individual participant the need to take control of one's own health. Through a workbook, the program helps individuals identify their own health risks caused by their personal lifestyle. Following that, the project helps participants to bring about intervention activities designed to change the part of their life or behavior that may be harmful to their health. Alcohol abuse is among the health problems which PLUS emphasizes. Individual counseling for each of these areas is provided, as are behavior change strategies.

The use of the occupational setting for prevention and/or health promotion activities is a promising development for future substance-abuse prevention efforts. A logical place to reach large numbers of adults is through work sites; in addition, the more complete or generic prevention efforts are more likely to be adopted by businesses. Finally, participants may also be reached and benefited more by this multifaceted health promotion strategy or prevention.

Drug Abuse and the Elderly

Another segment of the adult population yet to be adequately reached with substance-abuse* prevention programming is the elderly. Recent research has begun to describe the extent of the problem for this age group as well as the causative factors, and initial education efforts have been developed, focusing primarily on the prescription and over-the-counter (OTC) drug use by older Americans. The problem of alcohol abuse by the elderly is also being recognized, although prevention/education efforts on that topic are more limited. The elderly themselves are only a part of the problem sequence. Families and friends of the elderly; service providers, particularly those in the medical fields; and the manufacturing/distribution systems should each be participants in preventive efforts for older people.

The population over age 65 is growing at a rate three times faster than that of the general population. Today there are approximately 21 million older Americans, and projections indicate that there will be 27.5 million people over the age of 65 by 1990.[46] This older group suffers disproportionately from chronic health conditions and disabilities[47]; the frequency, extent, and intensity of physical disabilities are associated with age[48] and, therefore, a continuation of the elderly population increase will inevitably produce larger increases in the number of elderly with chronic illness and disability.[49]

There are no treatment measures that specifically modify many of these age-associated chronic illnesses,[50] but therapeutic agents are useful in relieving chronic disease symptoms. Older persons are, therefore, more likely to receive a greater quantity and variety of medications than are other age groups. For example, while the elderly consist of approximately 12% of the American population, they account for 25% of the prescription drug use in this country.[51] In addition, older adults may also self-medicate with over-the-counter drugs in an attempt to self-manage the health problems they face.

Research indicates that drug misuse or abuse is now a serious problem among the elderly population in America.[52-54] Cooper,[55] for example, in one study reported that 90% of the elderly have adverse reactions to the drugs they take, with 20% of these cases requiring

*For the purpose of this discussion only licit drug usage will be targeted, eg, the use of both prescription and over-the-counter medications.

hospitalization. These reactions of the elderly person to drugs are ones that often are unique to their age group; in addition, the physical changes that can occur in the elderly (physical deterioration, faulty memory, difficulty in reading fine print, etc) can also cause improper drug use in many cases. The average person over 60 consumes five drugs a day,[56] and researchers have found that they are often victims of polypharmacy or prescribing of multiple drugs,[57] frequently by several different physicians, thus increasing the opportunity for misuse. Iatrogenic disorders are not unusual in the older population, and adverse drug reactions are a serious problem with this group. Investigators claim that adverse drug reactions are responsible for from 3% to 5% of all hospital admissions[58] with a higher proportion seen among the aged. A 75-year-old person has a seven times greater chance of an adverse reaction than someone 25 years of age, and the incidence of side effects among the 70–79-year-old population is triple that of the 40–49-year-old group.[59] Unfortunately, an adverse drug reaction can go unrecognized because of its similarity to expected physical and mental dysfunction in the elderly.[60]

The extent of drug use and misuse has major economic as well as health importance for both the older adult and for society. Namey and Wilson[61] indicate that the elderly consume many more prescribed drugs than younger adults do, and Hammerman[62] reports that 10% of the older adult's health bill is spent purchasing drugs. Vestel[63] reported that the per capita expenditure by the elderly for prescribed drugs exceeds that for other age groups. In 1974 the elderly spent an estimated $2.3 billion on drugs, or more than 20% of the national total. Some of these costs, including the financial, eg, hospitalization costs for admissions resulting from adverse drug reactions, are paid for by society as well as by the individual.

The increase in health problems associated with aging also naturally brings about an increased use of medications, either physician- or self-prescribed and, again, an increased potential for abuse. The consumers' lack of necessary knowledge and understanding about drugs often makes them uninformed or poorly informed users; at the same time, many health care providers, including physicians and pharmacists, are not always knowledgeable about geriatric dosages, their patients' total drug history, or their understanding about proper usage procedures. Many of the various institutional health care givers, eg, nursing home and hospital personnel, also do not have an adequate knowledge and understanding of the proper use of drugs in the elderly and the subsequent potential for misuse.

Two additional factors to be considered in describing the extent of the problem are the producers, eg, the various pharmaceutical companies, and also the governmental bodies that regulate the manufacture and dispensing of drugs. Each of these also has a significant role in the entire medication process for the elderly. The pharmaceutical industry is, therefore, a part of the total problem of drug misuse among the elderly. As the older population, and public awareness of it, has expanded in the past quarter century, so has the industry's expansion of their elderly-oriented market, and its promotion.

Forms of Drug Misuse by the Consumer

Whittington et al[53] have classified drug misuse into four categories: overdose, underdose, erratic use, and contraindicated use. They state that:

> The patient who is defined as overusing drugs is one who either takes more than the prescribed amount of one drug or takes a pill [as needed] medication when it is not actually needed. Underuse, on the other hand, includes both the failure to acquire and take appropriate medications and the consistent failure to take as much of the drug as the prescription calls for. Erratic use refers to the patient's general failure to follow instructions and may include missed doses (underdose), double doses (over use), doses taken at the wrong time or by the wrong route of administration, and drug confusion, when the wrong drug is taken by mistake. Contra-indicated drug use occurs where the physician prescribes the wrong drug for his older patient—one which either is ineffective, produces unwanted or deleterious side effects, or interacts harmfully with other medications already being taken. (pp 1,2)

The older patient can experience any one or a combination of these types of misuse. Hoarding of medication is one situation that produces such drug misuse, and can be especially dangerous when the medicine is outdated. The elderly often fear that some day they will be cut off from their medication or will not be able to afford the cost so they save drugs for future use. Financial considerations are particularly important; a recent study[64] of reasons for noncompliance found the most frequently used excuse was the cost of the drugs. Sharing of medicine is also common, especially among those who live together in housing developments.[65]

Another important reason for medication error among older adults is one of attitude, long-ingrained values that influence how older adults feel about taking drugs. Some older individuals may be inveterate pill swallowers and others avoid drugs whenever they can. The fear of drug dependence may cause many individuals to discontinue drugs prematurely. Other factors that may be influential are the older adults' attitudes toward their health problem(s) and the sick role.

The elderly receive a disproportionate percentage of barbiturate sedatives and diazepam,[64] and are especially sensitive to the depressant qualities of barbiturates. Petersen and Whittington[66] found that 80% of drug reactions in the elderly "involved misuse of psychotropics (either a sedative or a tranquilizer), primarily Valium, Tuinal, and Luminal, and an additional 10% resulted from the misuse of the non-narcotic analgesic Darvon." The elderly have a reduced tolerance for these substances, and are, therefore, more at risk for reactions than are younger people.

A variety of aging-related factors combine to cause inadvertently incorrect use of various medications. It has been found, for example, that 15%–18% of the elderly have poor vision[67] and may, therefore, be simply misreading instructions. Poor memory also makes it difficult to remember the variety of instructions for taking numerous drugs. Often the person using more than one drug may become confused about when medication should be taken and in what quantity.[68] Schwartz and her associates[69] found that 47% of the time the patient omitted the medication (underuse), had inaccurate medication information (20%), or used incorrect self-medication (17%). Not following the physician's prescription is a common occurrence,[70] and the elderly are more likely than any other age group to fail to follow the orders of the physician.[69,71] This suggests not only a problem with the elderly in following directions, but also a lack of communication and understanding between patients and physicians and other health personnel.

Obviously, a major responsibility for proper drug use must be taken by the individual older person. Drug knowledge and attitudes are factors that influence these decisions, in addition to the previously described health related factors. Many people, for example, do not recognize their own responsibilities in the physician-patient relationship and the importance of gathering the data about their medications. Many times they are not sure which, or how many, medications they are ingesting, and a fear of bothering the physician prevents the communication that might alleviate this problem. In addition, the elderly person is often unaware of the types of information regarding drugs that the physician can provide and the importance of their fully understanding the reasons for their medication. Unfortunately, the rapid development of new drugs, lack of drug information and/or access to it, and inappropriate attitudes about substances limit the informed decisions that older individuals should make.

Each of these problem phases represents a causative factor in the underdose, overdose, erratic use, and contraindicated use classification. However, another significant cause is being recognized which produces a deliberate or uncaring misuse of drugs among the elderly. The psychosocial factor is increasingly seen as a reason for inappropriate drug use among older people. Pascarelli[72] reported an increasing number of elderly in treatment for drug dependence and abuse who cited isolation, poverty, low status, and physical disabilities, for example, as reasons for their inappropriate or unprescribed drug use. Reduced self-esteem and lack of satisfactory interpersonal relations caused by death, retirement, relocation, etc, are affective or psychosocial factors that must also be considered in this problem. Davison[73] found, however, that drugs are not the answer to many of the problems of the elderly. Instead, the person's lifestyle needs to be changed. Gaetano and Epstein[74] urge, therefore, that primary efforts be made with the elderly themselves, using a health education model, to change knowledge, attitudes, and behavior concerning drug use.

One program recently developed, under a grant from the Prevention Branch of the National Institute on Drug Abuse, is entitled *Elder-Ed: A Model Education Program, Prevention of Medication Misuse by the Elderly.* A three-section film addresses the problems of using both prescription and OTC drugs; patient-physician communication, purchasing drugs, and proper administration are included. The packet of written materials for distribution to older audiences includes a "Passport to Good Health Care" in which to record one's total medication history, a booklet on generic drugs, and a summary pamphlet. This project apparently parallels, with a new audience, traditional drug education program development; it is primarily cognitively based and does not refer to the affective or attitude components of drug misuse.

Gerontologists and agency staff who work with the elderly have begun to recognize the need for more prevention and/or education efforts, particularly those dealing with the psychosocial dimensions of abuse. A few senior centers and similar gerontology services are currently working with local prevention professionals to develop these approaches, and in the next few years it is likely that many new projects will be initiated.

Alcohol Abuse and the Elderly

Research also suggests that alcoholism and alcohol misuse are becoming a very serious problem among the elderly,[66,72] creating increasingly costly medical and social problems in this country. Simon et al[75] reported a 60% increase from 1950 to 1964 in the average annual death rate from alcoholic disorders (alcoholic psychosis, cirrhosis of the liver, and alcoholism) of all persons aged 20 years and over in the United States. In the older age groups the increase was 52% for white men aged 60–69, 27% for white men aged 70 and over, 114% for white women aged 60–69, and 39% for white women aged 70 and over. With the size of the elderly population expected to be much larger in the coming years, an increase can be expected in the number of people, especially women, who will enter old age with well-established drinking habits.[76]

However, very little alcohol research has been devoted specifically to the elderly despite the apparent magnitude of the problem. The initial findings do show that problem drinking among older people is a significant social health problem in all settings, including rural, suburban, and urban.[77] Recent reviews of alcohol problems of the elderly[78] indicate that between 2% and 10%, at least, of the general elderly population are affected. Traditional surveys have shown the highest evidence of alcoholism to be in the 34–50 age group. In Williams's[79] testimony before the Subcommittee on Alcoholism and Narcotics, and the Senate Subcommittee on Labor and Public Welfare, data from an informal survey of alcoholism information and referral centers were presented. The results indicated that one-third of the alcoholic or problem-drinker callers were people over the age of 55. She concluded that if this is representative of the total population of problem drinkers, then 3,000,000 of the 9,000,000 problem drinkers are aged over 55. There is a stigma attached to being aged or alcoholic in this country and the combination of the two is a serious problem.[80] While the elderly can be identified as a vulnerable sector of the adult population, insufficient research has been devoted specifically to describing the extent and causes of the problem, and the concept of prevention and early intervention among the elderly has not yet begun to be satisfactorily addressed.

The existing statistics regarding the incidence of alcoholism among the elderly may also be misleading, and in fact may only be the tip of the iceberg. Families try to protect the elderly alcohol abuser from being discovered to avoid embarrassment to the family.[81] For this reason the elderly person is hidden and untreated, and the habit becomes worse. Older problem drinkers also do not always appear in the "extent of the problem" statistics because they are often considered poor treatment risks and excluded from services and resources. Carruth et al[77] also found in their study that 30% of calls for alcohol referral and information services were for people over 55. Once intervention or treatment has started, however, the elderly alcoholic has met with remarkable success. Through the use of antidepressant medication and socialization programs the elderly alcohol abuser shows rapid progress toward recovery. All these people need is someone to listen to their problems and have a feeling that someone is concerned about their welfare. Once the feeling of depression has gone, the elderly find they no longer need alcohol to solve their problems.[82] Although the older alcoholic is more likely to drink daily, compared to their younger counterparts, they appear to have better response to treatment.

The problems of the elderly alcoholic are compounded by the mixture of alcohol with prescribed and/or over-the-counter drugs, an especially dangerous, even fatal, occurrence in some instances. This problem is more severe than previously thought.[83] Recognition of this syndrome is essential because there can often be intervention and/or cure. Again, medical and social agency personnel must be sensitized to the problem. In other settings, such as nursing homes, the problems of the older alcohol abuser are especially severe. There, this person tends to be a loner and often has trouble with interpersonal relationships. Care providers frequently have an unfavorable attitude toward older problem drinkers. This attitude is more negative than that shown toward the older person generally or toward the younger problem drinker. A more positive attitude is needed among social and health workers in order to improve the plight of the elderly alcohol abuser.

Causes of Alcoholism Among the Elderly

Two categories of elderly alcoholics have been identified.[84,85] The first includes alcoholics who have been drinking most of their lives and have survived the long-term effects of alcohol. Almost all of the persons who exhibit this pattern of alcoholism manifest some of the more serious conditions brought on by drinking, eg, cirrhosis of the liver and chronic brain syndrome.[83] The second type of elderly alcoholic, the one to whom prevention efforts should be directed, is the person who has begun drinking or abusing alcohol at an older age. These drinking problems are different from those of younger alcoholics. Drinking becomes a way of dealing with the stresses and identity problems that are associated with the aging process. Additional problems, such as the loss of a job or spouse, retirement, relocation, poor health, and boredom, may also lead to drinking.

Zimberg[82] points out that, "It is possible that the stresses of aging not only can lead to the development of late-onset alcoholism but also can contribute to the perpetuation of long-standing alcoholism." The incidence of alcohol abuse is thought to be higher among widowers and individuals with serious medical problems. Studies show that perhaps as many as 20% of elderly individuals who are medical inpatients, and 10% to 15% of elderly medical outpatients, have serious problems which are related to alcohol. Rosin and Glatt[85] found that elderly subjects with previously innocuous drinking behaviors listed three major influences that exacerbated their alcohol misuse. These related to physical, mental, and environmental aspects of aging, eg, bereavement, retirement, and loneliness.

A variety of developmental theories have described the psychosocial aspects of aging, all of which can contribute to later-onset alcohol abuse. Erikson[86] suggests that in the elderly an acceptance of one's life as having been meaningful can be difficult, especially the reconciliation of goals and accomplishments. Peck and Havighurst[87] believed that redefining one's worth and finding satisfaction in aspects of a person's self are important to happiness in older ages. Havighurst[88] recognized the difficulty of new learning that is necessary in later ages to face and solve new problems. Zusman[89] and Bengston[90] demonstrate with the "social breakdown syndrome" that an elderly person's sense of self, ability to mediate between self and society, and feelings of competence are linked with the social labeling and values experienced in aging. The syndrome "characterizes the dynamic relationship between susceptibility, negative labeling, and the development of psychological weakness.[90]

Prevention/Education Program Needs

One of the earlier-cited problems associated with many alcohol education programs for youth audiences has been the emphasis placed on information or cognitive materials, often without adequate consideration of the affective factors that contribute to the decisions regarding alcohol misuse or abuse. For an elderly target population with causative factors, such as loneliness, lack of self-esteem, fear, retirement, poor health, boredom, and bereavement, all of which have a significant affective psychosocial component, this domain, therefore, must be the primary focus of education efforts. The extent of the problem of the later onset of misuse or abuse of alcohol by older people indicates the need for prevention programs designed specifically for this audience and directed toward the factors that lead to the problem behavior.

While few alcohol abuse prevention programs have been developed for the elderly, one suggested approach involves the use of paraprofessional peer counselors. In a variety of mental health programs their help has been found to be less threatening to many older people. Butler and Lewis[91] report peer counselors provide new services in innovative ways, acting as role models and having enhanced understanding of how to help their cohorts. Peers can also be especially aware of characteristics unique to the group with which they are working. Their acceptance as leaders with their group is increased because of being "one of them." Paraprofessional peer counselors have a unique contribution to make to alcoholism programs (as AA proves). Peer counselors have frequently observed or experienced and resolved crises that lead to alcohol misuse or abuse in the elderly population. They may be especially adept in counseling others because they can bring their own personal experiences to the counseling situation. Waters et al[92] have pointed out that peer counselors have the advantages of having shared similar life experiences or at least having lived at the same time. They contend that this may lead to similarities in value structures which enhance effective communication with elderly clients. In conclusion, it appears that, while cognitive factors should not be ignored, affective modalities must play a major role in alcohol abuse prevention efforts with the elderly.

SUMMARY

Alcohol and drug abuse prevention programs must recognize that the final decision to abuse a substance rests with the individual. Even under the most extreme restrictions (eg, imprisonment), individuals are often able to use and/or abuse practically any substance. Therefore, a reasonable and realistic objective for prevention is to educate for responsible decision making. The foundations of responsible choice rest with a balance between substance-specific information and self-development.

Alcohol and drug abuse can occur at any age and in any setting. Future views of prevention must recognize this as well as the constantly evolving nature of human development, and the intrapersonal and interpersonal aspects as they change throughout the life span. As people mature, they have different needs for information about drugs and alcohol. Obviously, the pre-adolescent facing a critical stage in development needs a different kind of knowledge about drugs and alcohol than does the pre-retirement individual. Yet both need to reexamine their self-concepts and interpersonal relations. Drug and alcohol abuse prevention programs for clients of all ages must provide this age-appropriate balance between cognitive (drug- and alcohol-specific) and affective (development-stage-specific) components. In addition, prevention activities must also take a more generic approach, examining and sharing promising results from a variety of problem prevention fields, and coordinate developing new methodologies.

REFERENCES

1. Nowlis HH: *Commonalities Among Prevention Programs for Children and*

Youth. Paper presented at the First Annual Alcohol, Drug Abuse, and Mental Health Administration Conference on Prevention. Silver Spring, Md, September 1979, p 1.

2. Maloney SK: *Guide to Alcohol Programs for Youth,* National Institute on Alcohol Abuse and Alcoholism, US Government Printing Office, 1980. (Draft)

3. IMS America Ltd: *DAWN Quarterly Report,* Ambler, Pa, October-December 1978.

4. Swisher J, Hoffman A: Information: The irrelevant variable in drug education, in Corder B, Smith R, Swisher J (eds): *Drug Abuse Prevention: Strategies for Educators.* Dubuque, Iowa, WC Brown, 1974.

5. Braucht GN, Follingstad D, Brakarsh D, et al: Drug education: A review of goals, approaches and effectiveness. A paradigm for evaluation. *Q J Stud Alcohol* 34:1279-1292, 1973.

6. Zinberg N: *What is Drug Abuse?* Unpublished report, Drug Abuse Council, Washington, DC, 1976.

7. Blum RH: *Drug Education: Results and Recommendations.* Lexington, Mass, DC Heath and Company, 1976.

8. Ungerleider T: Drugs and the educational process. *Am Biol Teacher* 30:627-632, 1968.

9. Stuart RB: Teaching facts about drugs: Pushing or preventing? *J Educ Psychol* 6:189-201, 1974.

10. Goodstadt MS, Sheppard MA, Chan G, et al: *Drug Education Research: Review of the Past and Proposals for the Future.* Addictions Research Foundation, Toronto, Canada, 1979, p 18.

11. National Coordinating Council for Drug Education: *Drug Abuse Films.* Washington, DC, The National Coordinating Council for Drug Education and Information, 1973.

12. National Institute on Alcohol Abuse and Alcoholism: *Film Evaluations.* Rockville, Md, 1974.

13. Glaser F: Misinformation about drugs: A problem for drug abuse education. *Int J Addict* 5:595-609, 1970.

14. National Institute of Drug Abuse: *Toward a National Strategy for Primary Drug Abuse Prevention* (Final Report Delphi II). Rockville, Md, 1975.

15. Pennsylvania Governor's Council on Drug and Alcohol Abuse: *Annual Report.* Harrisburg, Penn, 1975.

16. *Life Skills for Mental Health,* Georgia Department of Human Resources, Division of Mental Health and Mental Retardation, Prevention Unit, Atlanta, 1977.

17. *Life Skills for Mental Health,* Focus on Mental Health, Division of Health, Safety and Physical Education, North Carolina Department of Public Instruction, Raleigh, 1976.

18. Mong ML: *Alcohol Specific Curricula: A selected list.* Rockville, Md, National Clearinghouse for Alcohol Information, 1980.

19. McManus M, Kassenbaum P: *CASPAR: A Prevention Prototype for School and Community Alcohol Education of Youth.* Rockville, Md, National Clearinghouse for Alcohol Information, 1977, p 11.

20. Kassenbaum P, McManus M: *Prevention X Three: Alcohol Education Models for Youth.* Rockville, Md, National Clearinghouse for Alcohol Information, 1978.

21. National Institute on Alcohol Abuse and Alcoholism: *The King County ESD Alcohol Education Curriculum.* Rockville, Md, 1978.

22. Carney RE: *An Evaluation of the Tempe, Arizona Drug Abuse Prevention*

Education Program. E.S.E.A. Title III Project, Grant No. 12-71-0021-0, Tempe Elementary School District #3, Tempe, Ariz, 1974.

23. Skiffington EW, Brown PM: Personal, home and school factors related to eleventh graders' drug attitudes. *Common Ground: Pennsylvania Prevention Forum,* June 1979, pp 2–9.

24. Visco EP, Finotti JF: Geomet Report Number HF-347. SPARK Program Analysis: Final Report. Gaithersburg, Md, Geomet, Inc, 1974.

25. Swisher J, Piniuk A: *An evaluation of Keystone Central School District's drug education program.* Harrisburg, Penn, The Pennsylvania Governor's Justice Commission, Region IV, 1973.

26. Swisher JD, Horan JJ: Effecting drug attitude change in college students via induced cognitive dissonance. *Journal of SPATE* 18:26–31, 1972.

27. D'Augelli J, Weener J: *Development of a parent education program as a drug abuse prevention strategy.* Final report to Governor's Council on Drug and Alcohol Abuse, June 1975.

28. Schaps E, DiBartolo R, Palley CS, et al: Primary prevention evaluation research: A review of 127 program evaluations. Pyramid Project, Walnut Creek, Calif, 1978.

29. ALMACA. An invitation—Membership Brochure. Reston, Va, Association of Labor-Management Administrators and Consultants on Alcoholism, 1977.

30. Von Wiegand RA: Alcoholism in industry (USA). *Br J Addict* 67:181–187, 1972.

31. Mandell W: *Does the Type of Treatment Make a Difference?* Paper presented to the American Medical Society on Alcoholism, 1971.

32. Trice H, Roman P: *Spirits and Demons at Work: Alcohol and Other Drugs on the Job.* Ithaca, NY, New York State School of Industrial and Labor Relations, Cornell University, 1972.

33. Urban M: Drugs and industry, in National Commission on Marijuana and Drug Abuse: *Drug Use in America,* vol I. Washington, DC: US Government Printing Office, 1972, pp 1136–1156.

34. Kurtis C: *Drug Abuse as Business Problem—The Problem Defined with Guidelines for Policy.* New York, New York Chamber of Commerce, August 1970.

35. Chambers C: *Differential Drug Use within the New York State Labor Force.* New York State Narcotic Addiction Control Commission, May 1971.

36. Rush H, Brown J: The drug problem in business: A survey of business and opinion and experience. *The Conference Board Record* vol 3, March 1971.

37. Levy S: A study of drug-related criminal behavior in business and industry. *Drug Forum* 3:363–372, 1974.

38. Korcok M, Seidler G (eds): *Focus on alcohol and drug issues* vol 1, February-March 1978.

39. Colson J: Alcohol awareness education in the US Air Force. *J Drug Education* 7:33–35, 1977.

40. Payne W, Monti J, Winer M: Industry as a setting for drug abuse education: A preliminary study. *J Drug Education* 6:201–208, 1976.

41. Martin J: The new business boom—employee fitness. *Nation's Business,* February 1978.

42. *Time.* From boardroom to locker room. New York, Time, Inc, January 22, 1979.

43. Warner R: Boeing company fitness programs involve calisthenics, running, *AAFDBI Action* vol 1, July 1978.

44. Martin J: Corporate health: A result of employee fitness. *The Physician and Sportsmedicine.* March 1978.
45. Kansas Department of Health and Environment, *PLUS,* Topeka, Kan, 1979.
46. US Bureau of the Census: Illustrative population projections for the US: The demographic effects of alternate paths to zero growth. *Current Population Reports* (Series P-25, No. 480), 1972.
47. Atchley R: *The social forces in later life. An introduction to social gerontology,* ed 2. Belmont, Calif, Wadsworth, 1977.
48. Dovenmuekle R, Busse E, Newman G: Physical problems of older people, in Palmore E (ed): *Normal aging: Reports of the Duke Longitudinal Study.* Durham, NC, Duke University Press, 1970.
49. Lillienfeld A, Gifford A: *Chronic disease and public health.* Baltimore, Johns Hopkins University Press, 1966.
50. Hylbert K, Hylbert K: *Medical Information for Human Service Workers.* State College, Pa, Counselor Education Press, 1977.
51. Basen MM: The elderly and drugs: Problem overview and program strategy. *Pub Health Rep* 92:43–48, 1977.
52. Heller FJ, Wynne R: Drug Misuse by the Elderly: Indications and Treatment Suggestions, in Senay E, Shorty V, Alksne H (eds): *Development in the Field of Drug Abuse.* Cambridge, Mass, Schenkman Publishers, 1975, pp 945–955.
53. Whittington FJ, Petersen DM, Beer ET: *Drug Misuse Among Older People.* Paper presented at the Annual Meeting of the Gerontological Society, Dallas, Tex, November 16–20, 1978.
54. Pascarelli EF, Fischer W: Drug dependence in the elderly. *Int J Aging Hum Dev* 5:347–356, 1975.
55. Cooper JW: Implications of drug reactions—recognition, incidence and prevention. *RI Med J* 58:274–356, 1975.
56. Hagar M: Giving medication to the elderly a hazardous practice. *The US Journal* July 1977, p 8.
57. Lamy PP: Geriatric drug therapy. *Clin Med* 81:52–57, 1974.
58. Caranasos G Jr, Stewart RB, Cluff LE: Drug induced illness leading to hospitalization. *JAMA* 228:713–717, 1974.
59. Center for Human Services: *Wise Drug Use for the Aging: The Role of the Service Provider.* Washington, DC, Pilot test copy.
60. Kayne R: Drugs and the aged, in Burnside I (ed): *Nursing and the Aged.* New York, McGraw-Hill, 1976.
61. Namey C, Wilson R: Age patterns in medical care, illness, and disability. *Vital and Health Statistics* 10(70), 1972.
62. Hammerman J: Health services: Their success and failure in reaching older adults. *Am J Public Health* 64:253–256, 1974.
63. Vestel RE: *Drugs and the Elderly.* National Institute on Aging Science Writer Seminar Series, 1976.
64. Brand FN, Smith RT, Brand PA: Effect of economic barriers to medical care on patients' non-compliance. *Pub Health Rep* 92:72–78, January-February 1977.
65. Pascarelli EF: Drug dependence: An age old problem compounded by old age. *Geriatrics* 29:109–115, 1974.
66. Petersen DM, Whittington ES: Drug use among the elderly: A review. *J Psychedelic Drugs* 1:25–37, 1977.
67. Kornzweig AL: The eye in old age. In Rossman I (ed): *Clinical Geriatrics.* Philadelphia, JB Lippincott, 1971, pp 229–246.

68. Davis JM, Fann WE, Khaled El-Yousef M, et al: Clinical problems in treating the aged with psychotropic drugs, in Eisenoreer C, France WE (eds): *Psychopharmacology and Aging.* New York, Plenum Press, 1973.

69. Schwartz D, Wang M, Zietz L, et al: Medication errors made by chronically ill elderly patients. *Am J Pub Health* 52:2018–2059, 1962.

70. Warren F: Self medication problems among the elderly, in Petersen DM, et al (eds): *Drugs and the Elderly: Social and Pharmacologic Issues.* Springfield, Ill, Charles C Thomas, 1979.

71. Brand FN, Smith RT, Brand PA: Effect of economic barriers to medical care on patients' non-compliance, *Pub Health Rep* 92:72–78, 1977.

72. Pascarelli EF: Drug dependence: An age old problem compound by old age. *Geriatrics* 29:109–115, 1974.

73. Davison W: Pitfalls to avoid in prescribing drugs for the elderly. *Geriatrics* 30:152–158, 1975.

74. Gaetano RJ, Epstein BT: Strategies and techniques for drug education among the elderly, in Petersen D, Whittington F, Payne B (eds): *Drugs and the Elderly.* Springfield, Ill, Charles C Thomas, 1979.

75. Simon A, Epstein LS, Reynolds L: Alcoholism in the geriatric mentally ill. *Geriatrics* 23:125–131, 1968.

76. US Department of Health, Education and Welfare. Second special report to the US Congress on alcohol and health, 1974.

77. Carruth B, Williams EP, Mysack P, et al: Community care providers and the problem drinker. *Grassroots,* July suppl, 1975, pp 1–5.

78. Schuckit MA, Miller PL: Alcoholism in elderly men: A survey of a general medical ward. *Ann NY Acad Sci* 273:558–571, 1975.

79. Williams EP: Testimony before the subcommittee on alcoholism and narcotics and the subcommittee on labor and public welfare, in *Alcohol and Drug Abuse Among the Elderly.* Washington, DC, US Government Printing Office, June 1976.

80. Snyder V: Aging, alcoholism and reactions to loss. *Social Work* 5:232–233, 1977.

81. Droller H: Some aspects of alcoholism in the elderly. *Lancet* 2:137–139, 1964.

82. Zimberg S: The elderly alcoholic. *Gerontologist* 14:221–224, 1974.

83. Pascarelli EF, Fischer W: Drug dependence in the elderly. *Int J Aging Hum Dev* 5:347–356, 1974.

84. Zimberg S: The geriatric alcoholic on a psychiatric couch. *Geriatric Focus* 11:6–7, 1972.

85. Rosin AJ, Glatt MM: Alcohol excess in the elderly. *Q J Stud Alcohol* 32:53–59, 1971.

86. Erikson EH: *Childhood and Society.* New York, WW Norton, 1950.

87. Peck RF, Havighurst RJ: *The Psychology of Character Development.* New York, John Wiley & Sons, 1960.

88. Havighurst R: Social and psychological aspects of aging. In Gorlow L, Katkovsky W (eds): *Readings in the Psychology of Adjustment.* New York, McGraw-Hill, 1959.

89. Zusman J: Some explanations of the changing appearance of psychotic patients: Antecedents of the social breakdown syndrome concept. *Milbank Mem Fund Q* 64:2, 1976.

90. Bengston V: *The Social Psychology of Aging.* New York, Bobbs-Merrill, 1973.

91. Butler RN, Lewis MI: *Aging and Mental Health.* St. Louis, CV Mosby, 1977.

92. Waters E, Reiter S, White B, et al: The role of paraprofessional peer counselors in working with older people, in Ganikos ML, Grady LA (eds): *Counseling the Aged: A Training Syllabus for Educators.* Washington, DC, American Personnel and Guidance Association, 1979.

8 Research Relating to Alcohol and Opiate Dependence

Edward Gottheil
Bradley D. Evans
Karl Verebey

Historically, alcoholism and narcotic addiction have generally been considered to be separate disorders. It was considered inappropriate to treat alcoholic and narcotic addicts in the same clinical setting because it was believed that alcoholics were different from narcotic addicts. The former appeared to be older, more dependent, and more neurotic than their sociopathic narcotic addict counterparts. We now know that both groups of patients have much in common, and patients abusing both alcohol and the opiates as well as other substances (the polydrug abusers) are being seen with ever increasing frequency.

Attempts to sort out the differences and similarities between alcohol and narcotic abusers have been made from a variety of points of view. Psychologists, for example, have searched for distinguishing personality characteristics; sociologists have attempted to understand the background factors predisposing to the development of the various addictions; behaviorists have explored the role of conditioning; geneticists have looked for possible inherited links between alcoholism and narcotic abuse as well as relationships to affective illness and sociopathy; and biochemical investigators have focused on possible neurochemical explanations for addictive vulnerability. The emerging findings suggest that it may be more profitable to view alcoholism and narcotic addition as similar rather than as separate disorders. Accordingly, we shall review findings that emphasize the similarities rather than the differences between alcohol and drug abuse. Our review of developmental, psychometric, genetic, and biochemical and pharmacological studies, therefore, will stress this point of view.

DEVELOPMENTAL ASPECTS

Many investigators have studied adolescents in attempts to find particular developmental patterns and differentiating features of individuals with a predisposition to addiction. Their underlying assumptions were that, if such patterns and features could be isolated, they would likely have important implications for etiology and be helpful in diagnosis. Furthermore, they should assist us in the early identification of vulnerable individuals and possibly suggest more appropriate treatment and prevention strategies.

McCord et al,[1] in a well-known, longitudinal study of boys seen at a child guidance clinic, noted several factors that were predictive of later alcoholism. These included maternal alternation between affection and rejection; maternal escapism; maternal deviance, such as criminal activity and alcoholism; maternal denigration of the father; antagonism between the parents; and maternal resentment of her role in the family. A frustrating and depriving background is also related to the development of alcoholism in women and contributes to dependency, self-depreciation, insecurity in the sexual role, and feelings of inadequacy.[2] In addition, early parental loss may be especially important in the development of female alcoholism. Of the 451 alcoholic women studied by DeLint,[2] 166 (35%) had lost one or both parents before the age of 16, as compared to only 264 of 2005 (13%) alcoholic men. In her review of the literature, Lindbeck[3] concluded that alcoholism in women is related to a poor self-image, feelings of inadequacy and insecurity in the feminine role, and sensitization to loss.

Demone's[4] survey of adolescent drinking practices indicated that most adolescents drink, and their drinking behavior largely reflects that of their parents and significant others in their environment. Abstainers and pathological drinkers both were removed from the mainstream of adolescent activity. The pathological drinkers (7% of the sample by age 18) were generally unhappy individuals whose homelife had been inadequate. They had parents who either drank significantly or were abstainers. They could not confide in their parents and were more dependent on peer relationships. Truancy, one or more grades repeated, and an unwillingness to participate in adult-sanctioned activities were characteristic. When they drank, they usually drank to drunkenness, committed antisocial acts, and had difficulties with the police.

Jessor and Jessor[5] studied personality, social, and behavioral variables that would be expected to predict the development of problem drinking based on social learning theory.[6,7] They collected data on over 2000 junior high school students whom they classified as abstainers, drinkers without problems, and problem drinkers.

Problem drinkers were defined as those with a history of public drunkenness and problems relating to friends, school, dating, police, driving, and the family. In comparison to the non-problem drinkers, the problem drinkers: exhibited more aggression, stealing and lying; were more involved with other substances of abuse, notably marijuana; engaged in more sexual activity; placed less value on achievement and more value on independence; expected less achievement and had lower grade point averages; were more tolerant of social deviance; and perceived more peer support for drinking and drug use.

The Jessors then followed the group of non-problem drinkers for a one-year period. As compared to those who continued to drink without developing significant problems, those who became problem drinkers one year later had previously exhibited more delinquent behavior, expected less achievement, perceived more peer support for drug use, and more frequently drank to the point of drunkenness while not differing in the total amount of alcohol ingested. These latter variables, then, were not only characteristic of adolescent problem drinkers, but they existed prior to and were predictive of problem drinking.

In a study of junior and senior high school students, Hamburg et al[8] found that the attitudinal and behavioral correlates of adolescent drug use were similar to those of adolescent problem drinkers. Students with higher levels of drug use had poorer relationships with their parents, spent more time at raps and parties than at sports and homework, were less active in religion, had vaguer ideas about their futures, were more knowledgeable about and accepting of drugs, had poorer relationships to teachers, were more critical of drug education programs, and had had lower grade point averages. These characteristics became more pronounced with increasing levels of use and abuse. Thus, Ausubel[9] described young male addicts as immature, inadequate, passive-aggressive individuals for whom narcotics have a special appeal; and Gilbert and Lombardi[10] found that the main distinguishing features of young institutionalized male addicts were psychopathy, depression, tension, insecurity, feelings of inadequacy, and difficulty in forming close relationships.

In their review of the causes of drug abuse, Glasscote et al[11] summarized the characteristics found in most drug abusers as problems in socialization; low thresholds for frustration, disappointment and pain; need for immediate gratification; dependency; sexual immaturity; poor inner controls; and difficulty in coping with life and society. They believed that many are sociopathic personalities. Chinlund[12] reported that female addicts also display similar sociopathic personality patterns.

In addition to the similarities noted between young male and female users and abusers of alcohol and other drugs, the surveys described also indicate a relationship between alcohol and drug use. Instead of being restricted to one or the other, the same individuals tend to use or abuse both. Wechsler and Thum's[13] study of 700 students in grades 6 through 12 in Town B is a case in point. They classified the students into five categories of increasing alcohol intake (abstainer to relatively frequent user with drunkenness) and found that the proportion of marijuana users increased across the five drinking categories from 1% through 24%, 21%, 66%, and 78% respectively. Similarly, barbiturate use increased from 1% through 6%, 5%, 14%, and 40%; and amphetamine use from 1% through 2%, 5%, 15%, and 32%. Increased use of one substance seemed to predispose, then, to increased rather than decreased use of other substances.

Westermeyer and Walzer[14] evaluated the extent of alcohol and other drugs use in 100 consecutive patients, between the ages of 15 and 25, admitted to a general psychiatric unit of a university hospital. Their findings indicated that, in this population, alcohol and drug use were not related to demographic and clinical variables, but were related to social variables, such as felonies, truancy, running away, pregnancy out of wedlock, divorce, separation, and vehicular accidents while intoxicated. These correlations occurred, moreover, throughout the range of diagnostic categories. It could be concluded only that heavy substance use was correlated with sociopathic behavior, irrespective of the primary psychiatric diagnosis.

Impulsiveness, restlessness, and aggressiveness, in addition to being characteristic of young alcohol and drug users and abusers, are symptoms that today are also considered typical of individuals with minimal brain dysfunction (MBD). Renshaw[15] discussed hyperactivity in children as a syndrome or collection of behaviors which include shortened attention span, extreme excitability, poor impulse control, impaired concentration, destructiveness, and disturbed interpersonal relationships. Hyperactive children most often are discovered by their school teachers because of their disruptive behavior between the ages of five and seven. The condition is approximately four times as common in boys than girls. Mendelson et al[16] found that 15% of hyperactive children were already excessive users of alcohol between the ages of 12 and 16. Morrison and Stewart[17-19] reported that hyperactive children were more likely than controls to have parents who were alcoholic, suggesting a possible genetic link between alcoholism and minimal brain dysfunction. Goodwin et al[20] found that alcoholics more frequently had symptoms of hyperactivity as children when studied retrospectively. Tarter[21] advanced the hypothesis that at least a subgroup of alcoholics existed that suffered from hyperactivity and other soft neurological signs indicative of minimal brain dysfunction prior to the onset of drinking. He suggested that neuropsychological deficits in such individuals might play a crucial role in the etiology and maintenance of their alcoholic condition.

Severe alcoholics (primary), less severe alcoholics (secondary), psychiatric controls, and normal controls were compared with respect to scores on an MBD checklist of symptoms and ratings of familial drinking.[22] The primary alcoholics were found to endorse significantly more of the childhood symptoms of MBD than the secondary alcoholics, and both alcoholic groups reported more alcoholic relatives than the nonalcoholic control groups. It may be, then, that certain individuals inherit a vulnerability that predisposes to MBD and to alcoholism. The findings, of course, need to be replicated and, in view of the possible linkages between alcoholism and drug addiction, it would be most interesting to extend these studies to include drug addicts.

Even this brief review seems to indicate that male and female alcoholics and drug addicts emerge from similarly frustrating and depriving backgrounds where parental antagonism, drinking, antisocial behavior and values, and broken homes are common. Given this background, developmental, learning, social, and psychodynamic theories[23] are invoked to explain the not unexpected findings that users and abusers are characterized by low self-esteem, dependency, hostility, impulsiveness, self-depreciation, insecurity in the sexual role, feelings of inadequacy, and social ineptness. There is a turning away from the family and a search for peer support and acceptance. Low expectations for achievement and acceptance lead to poor grades, association with similar peers, delinquent behavior, drinking, and drug use. The use of addictive substances provides escape, gratification, and social group belonging.

The background and adolescent characteristics of the alcoholic and drug addict are not specific. They do not permit us to differentiate one from the other, nor do they differ from the sociopath. There is also considerable overlap with depression in terms of early parental loss (especially in the female), low self-esteem, self-depreciation, and insecurity.

Adolescents who use more alcohol also use more drugs and exhibit the symptoms of impulsiveness, restlessness, and aggressiveness which are reminiscent of MBD. It may be that there are certain individuals who "self-medicate" with alcohol or opiates for such an underlying neurobiological problem or defect that is genetically transmitted. If the finding that there is a subgroup of alcoholics who inherit a vulnerability to MBD, and severe alcoholism is replicated and extended to include a subgroup of drug addicts, an interesting biological explanation of addictive disease may take its place alongside of the more usual psychosocial explanations.

PSYCHOMETRIC STUDIES

Despite differing explanations for the development of alcoholism and drug abuse, many investigators believe that addiction represents some type of underlying emotional disturbance. Psychometric studies have attempted to identify and describe such disturbances and have sought to determine characteristic personality profiles of alcoholics and drug addicts, and to differentiate them from each other as well as from other pathological groups and normal controls.

The psychoanalysts provided many of the early personality descriptions of alcoholics and drug addicts. Knight,[24] for example, described in the alcoholic a pattern of dependency, rejection, and an insatiable desire for indulgence, associated with feelings of inferiority and guilt. Menninger[25] pointed to a self-destructive drive as the prime component in alcoholism. The hostility is expressed through antisocial behavior while intoxicated, and by guilt and depression when sober. Fenichel[26] attributed oral and narcissistic traits to narcotic addicts as well as alcoholics, while Rado[27] did not separate drug and alcohol abusers. He considered all drug craving as a manifestation of a single disease which he called pharmacothymia. He believed that certain individuals had the potential to respond actively to the pleasurable effects of drugs, and that it was not the drug per se but the impulse to use it that made certain individuals use drugs in response to frustration. Knight and Prout[28] found that narcotic addicts have Rorschach protocols indicating inadequate personalities, motivated by immediate needs and the attainment of immediate goals, and Thematic Apperception Test records indicating insecurity and an unwillingness to comply with authority.

Brill,[29] on the basis of case history and interview data, described post-hospitalization narcotic addicts in terms of dependence, insecurity, feelings of inadequacy, low frustration tolerance, confusion in the sexual role and identity, poor interpersonal relationships, and hostility. Hartman[30] found drug addicts to be orally fixated, and Sharoff[31] suggested that they withdraw from conflict to avoid having to struggle with or experience self-condemnation. These dynamic descriptions of the alcoholic and narcotic addict are very similar to each other and to those reported earlier in the studies of adolescent alcohol and drug users and abusers.

Reith et al[32] used the Edwards Personality Preference Schedule to compare heroin addict and nonaddict prisoners. Seventy pairs of addicted and nonaddicted criminal offenders were matched for age, educational level, intelligence, and home environment. The addicted offenders showed greater aggression and dependency needs than the nonaddicted criminals, suggesting that the addicted group had an inappropriate, if not conflicting, approach to expressing their aggressive drives. The addicted offenders also showed less endurance or willingness to complete tasks so that they appeared more impulsive and less persistent than nonaddicted offenders. Conflicts over aggression, dependence, and impulsivity were previously noted as prominent characteristics in the psychodynamic descriptions of narcotic addicts and alcoholics.

The Minnesota Multiphasic Personality Inventory (MMPI) has been the most frequently used, objectively scored instrument in comparisons of alcoholics and drug addicts. The findings, however, differ from study to study. Some of these differences reflect population differences, but some are not as readily explained. Hill et al[33] used the MMPI to compare white male alcoholics, narcotic addicts, and criminals, and concluded that undifferentiated psychopathy was the common personality characteristic of the three populations. Heller and Mordkoff[34] administered the MMPI to 67 male and female, young, middle-class, nonaddicted polydrug abusers at a nonresidential drug rehabilitation center. Their results showed elevated scores on psychopathic deviance *(Pd)*, and on schizophrenia *(Sc)*, depression *(D)*, and psychasthenia *(Pt)*, similar to the results of Gilbert and Lombardi,[10] but different from those of Hill et al[33] and Olson[35] whose older, addicted patients had elevated scores in the areas of psychopathic deviance *(Pd)* and mania *(Ma)*. Heller and Mordkoff[34] suggested that there were two basic groups of drug abusers: antisocial types with little evidence of overt reports of anxiety, guilt, insecurity, or depression (addicts); and those characterized by antisocial tendencies but evidencing signs of anxiety and depression (nonaddicted, multiple drug abusers).

In an effort to more carefully delineate differences and similarities between the MMPI profiles of alcoholics and narcotic addicts,

Overall[36] used the MMPI data of alcoholics and narcotic addicts collected by Lanyon.[37] The scores of 640 patients were subjected to discriminant function analysis to evaluate the nature and statistical significance of differences in the MMPI profiles of alcoholics and narcotic addicts. Both groups were similar in showing elevated *Pd* scores, but alcoholics, like the nonaddicted polydrug abusers of Heller and Mordkoff,[34] also had elevated scores on *D, Hy, Pt,* and *Sc.* Overall concluded that profiles in which *Pd* and *Ma* are elevated relative to *Hy* and *Pt* are more likely associated with narcotic addiction, whereas alcohol abuse is suggested by elevated scores on *Hy* and *Pt* in addition to *Pd* and *Ma.* Nevertheless, Gilberstadt and Duker[38] suggested that the *Pd-Ma* personality pattern, essentially Overall's narcotic addict prototype, indicates a proneness to heavy drinking, and Goldstein and Linden[39] described a subgroup of alcoholics that was characterized by the *Pd-Ma* MMPI pattern. Moreover, Weissman et al[40] and Senay[41] have presented data showing that depression as measured by various rating scales is more common in narcotic addiction than is suggested by Overall's study, occurring in approximately one-third of methadone-maintained patients.

Apfeldorf,[42] after his review of the various alcoholism scales derived from the MMPI, suggested that the MacAndrew Scale[43] is the most effective for distinguishing alcoholics from other psychiatric groups and controls. However, when Kranitz[44] attempted to study the problem of differentiation between different drug-abusing populations by administering the MacAndrew Scale to heroin addicts as well as to hospitalized alcoholic and psychiatric patients, he found that the scale did not differentiate between alcoholics and heroin addicts. It would appear, then, that the scale may not be measuring specific characteristics of alcoholics but characteristics common to all drug-dependent patients.

Herl[45] used a variety of tests, including the Jessor Alienation Scale, the Rokeach Dogmatism Scale, the Rotter Internal-External Locus of Control Scale, and the Zuckerman Sensation Seeking Scale to compare 75 psychiatric drug abusers, 75 psychiatric nonabusers, and 75 controls. He found the groups similar on all measures except that the drug abusers showed a greater preferred level of self-stimulation or sensation-oriented experiences than nonabusers. Zuckerman found that sensation seeking was associated with alcohol and drug use as well with hypomanic tendencies, delinquent behavior, and sociopathy.[46–48] Sensation seeking has also been found to be related to decreased platelet monoamine oxidase activity, a physiologically important biochemical feature of chronic schizophrenia and alcoholism.[49,50] Furthermore, there is some evidence for a genetic basis for sensation seeking.[48] Thus, Zuckerman's scale seems to provide some interesting links between the psychology and the biology of substance abuse.

Psychodynamic and psychometric descriptions of alcoholics and drug addicts are very similar to each other and to the descriptions of adolescent alcohol and drug users and abusers. While some studies indicate that alcoholics, polydrug abusers, and younger users tend to show more neurotic features than narcotic addicts, by and large, it has been most difficult, using a variety of objective psychometric tests, to differentiate alcoholics, narcotic addicts, and even nonaddicted criminals. There is significant overlap in the profiles of alcoholics and drug addicts, especially in the areas of sociopathy and depression. Relationships between sensation seeking and alcohol use, drug use, hypomania, sociopathy, and decreased platelet monoamine oxidase activity suggest interesting possible links between the psychology and biology of substance abuse.

GENETIC ASPECTS

The most useful methods for the investigation of genetic factors in psychiatric disorders have been family, twin, and adoption studies. Family studies provide reasonable risk estimates among different classes of relatives of affected patients. These are obtained from the frequency of occurrence of a particular illness in the various categories of family members. For example, Rosenthal[51] reported that in families of male alcoholics, the risk estimate for alcoholism in mothers was 0.4% to 8.2%, and in sisters 2.8% to 10.3%. The risk for male relatives was roughly five times higher than that for females: 11.4% to 32.6% for fathers, and 12.2% to 27.7% for brothers. In comparison, a survey of studies by Goodwin[52] indicated risk rates for alcoholism in the general population to be 3%–5% for males and 0.1%–1% for females. In a review of 39 studies reporting on the families of 6251 alcoholics and 4083 nonalcoholics, Cotton[53] found that the frequency of alcoholism in fathers of alcoholics ranged from 2.5% to 50% with an average of 27%, and the frequency in mothers ranged from 5% or less to 27%, with an average of 4.9%. The frequency of alcoholism in fathers of nonpsychiatric, nonalcoholics averaged 5.2%, and in mothers 1.2%. Another interesting finding in the Cotton paper was that women alcoholics are more likely than men alcoholics to come from families in which there was parental alcoholism.

The predominance of males in samples of alcoholics has led to the suspicion that there may be an X-linked genetic transmission of alcoholism. If true, it would be expected that sons of daughters of alcoholics would show a higher incidence of alcoholism than the sons of sons of alcoholics. Kaij and Dock,[54] however, found no evidence to support this hypothesis. In their family studies, the rate of alcoholism for all grandchildren of alcoholics was 43%. Such high rates would, if anything, suggest a dominant rather than a recessive form of transmission.

While family studies provide risk estimates that are useful for planning prevention and treatment strategies and resources, they do not serve to adequately distinguish environmental from genetic factors. Twin studies involve a comparison of monozygotic (MZ) and dyzygotic (DZ) twin pairs. The monozygotic twins are considered to be 100% genetically identical as compared to 50% for the dyzygotic twins. If environmental effects for both types of twin pairs are assumed to be equal, higher MZ than DZ concordance rates would then be indicative of genetic influence. For example, in a well-designed twin study, Kaij[55] examined concordance rates according to the severity of alcohol abuse. For the entire sample, the DZ concordance rate was 31.5% and the MZ rate was 54.2%. In the categories of greatest severity, DZ twins showed a high concordance rate of 66.7%; however, the MZ concordance rate of 84.5% was still significantly higher.

Adoption studies are concerned with persons separated from their biological parents soon after birth and raised by biologically unrelated parents. If the adoptee is significantly more like his biological relatives than his adoptive relatives, genetic factors are assumed to be present. In the Collaborative Danish-American Study,[56,57] the investigators identified 55 adoptees who had at least one biological parent with alcoholism severe enough to require hospitalization. These children were adopted before six weeks of age by nonalcoholic foster parents, and, further the children had no known contact with their biological parents. There were 78 controls who met the same adoption criteria but had no alcoholic biological parent. When the adoptees were examined without knowledge of their backgrounds, the incidence of alcoholism in the proband group was found to be 18%, nearly four times higher than the 5% incidence in the control group.

Bohman[58] compared 50 male and 50 female adoptees whose biological fathers were hospitalized for alcoholism with control adoptees who were carefully matched for age, sex, age at placement, occupations of adoptive parents, and ages of biological and adoptive parents at the time of the child's birth. Of the males whose fathers were alcoholic, 20% were also alcoholic in comparison to 6% of the controls. The number of adopted female alcoholics was too small for statistical study. These results are essentially identical to those reported by Goodwin.[56,57,59]

There is little direct evidence to indicate that narcotic addiction is genetically transmitted. Nevertheless, a number of studies suggest interesting relationships and possible genetic links among alcoholism, drug addiction, affective illness, and sociopathy. Feighner et al,[60] for example, found that family members of alcoholics show an increased incidence of affective disorders. Studies by Schuckit et al[61] and Winocur et al[62] revealed an increase in both affective disorders and sociopathy in families of alcoholics. In Cotton's[53] review, relatives of alcoholics as compared to relatives of controls had higher rates of alcoholism; affective illness, especially depression; and sociopathy.

Clinically, Pottenger et al[63] reported that a substantial number of alcoholics who seek help for their alcoholism also suffer from persistent and severe depression. Another similarity between alcoholism and depression is found in the periodicity of both illnesses. Discrete episodes of illness followed by periods of relative calm are commonly seen in affective disorder; in alcoholism, there is "binge drinking" and the "bender."[64] In addition, suicide rates are high in alcoholism and affective disorder. The life-long risk of suicide in alcoholics is estimated at 15%–20%,[65] which is similar to the risk of 15% reported for affective illness.[66]

Depression also seems to be relatively common in opiate-dependent individuals. Weissman et al[40] and Senay[41] have shown that significant depression, as indicated by various rating scales, occurs in approximately one-third of patients in methadone treatment. The course of drug addiction also progresses episodically from brief and less frequent periods of use and abuse to more frequent periods of increasing duration.

With respect to suicide, it is often difficult to estimate what proportions of drug overdose deaths are due to accident or to suicide, but the suicide risk in depressed addicts is certainly considerable.

Finally, Maddox and Elliott[67] found that when alcoholism and opiate dependence are combined, the extent of depression is increased even further. Since alcoholism and drug addiction share many features of affective illness, and since there is considerable evidence in both twin studies[68-70] and family studies[71-74] of the genetic transmission of affective illness, it is conceivable that some biological link might exist between alcoholism and opiate dependence.

Developmental and psychometric studies have suggested that alcoholics and narcotic addicts also have many sociopathic traits in common. While criminality per se may not be genetic, some studies suggest that sociopathy might be.[58] For example, some twin studies have shown monozygotic concordance rates for sociopathy to be higher than dyzygotic rates.[75,76] Adoptee studies have found antisocial personalities in 13% of adopted away children of female legal offenders as compared to only 2.3% for all other adoptees.[77-78] A significant correlation is reported between antisocial personality in adoptees and their biological parents.[79] On the basis of their evidence, Cloninger et al[80] have suggested that there may be a multifactoral genetic transmission of sociopathy.

In summary, family, twin, and adoptee studies have provided evidence for the genetic transmission of alcoholism, depression, and sociopathy. While there is no clear evidence reported for the genetic transmission of opiate dependence, the possibility of a link between alcoholism and opiate dependence is suggested by the fact that depression and sociopathy are common to both groups. The argument is certainly not conclusive, since two conditions A, and B, may be related to an overlap with C and still be independent of each other. Nevertheless, given the genetic vulnerability of sociopathy, depression, and alcoholism, it is possible that a similar vulnerability exists for opiate dependence. This possibility would seem to be worth exploring. If such a vulnerability were found, it might suggest a *common biochemical link* between alcoholism and opiate dependence, predict "at-risk" individuals, and lead to different prevention and treatment strategies.

BIOCHEMICAL AND PHARMACOLOGICAL STUDIES

The developmental, psychometric, and genetic studies that were reviewed have pointed to many similarities between alcohol- and opiate-dependent individuals, thereby suggesting the possibility of common biological mechanisms and genetic predispositions. Recent psychobiological studies, especially those involving monoamine oxidase, the isoquinoline alkaloids, and the endorphins, provide further evidence of possible links between the psychology and biology of the addictions.

Monoamine Oxidase

Monoamine oxidase (MAO) activity has been studied in a variety of psychiatric conditions. The importance of this enzyme derives from its involvement in the metabolic breakdown of crucial amine neurotransmitters through the process of oxidative deamination. Moreover, its occurrence in the platelet, which is considered to have physiochemical properties similar to those of nervous tissue,[81] makes it easily accessible for study.

Platelet MAO activity has been found to be reduced in patients with bipolar affective illness and also in the relatives of these patients.[49,82] It is also reduced in chronic psychotic states such as schizophrenia.

In alcoholics, enzyme activity was found to be low during acute intoxication and withdrawal.[83,84] Sullivan et al[85] reported that platelet

MAO activity was significantly lower in alcoholics than in controls at 3, 6, and 12 months after discharge. MAO activity has also been reported to be low in the brains of alcoholics who had committed suicide,[86] although this was not found by Grote et al.[87]

In general, the evidence seems to suggest that the enzyme may be an inheritable and clinically stable characteristic,[88,89] the activity of which is reduced in alcoholism and in other psychiatric conditions. We previously noted that alcoholism, drug abuse, affective illness, and sociopathy demonstrate a number of interesting relationships with respect to background, personality, and genetic factors. In addition, sensation seeking was found to be associated with these conditions and also with MAO activity. Although much research remains to be done, the findings thus far suggest the intriguing speculation that MAO may be involved in a biochemical system that is the mechanism for some of these relationships. However, we need more systematic information about the factors that increase or decrease enzyme activity in both acute and chronic situations. MAO activity levels in young drinkers, drug users, drug abusers, and sociopaths have not yet been determined. Family, twin, and adoptee studies are also needed to more clearly demonstrate the genetic transmission of MAO activity levels and evaluate whether such transmission carries with it a risk to develop particular behavior patterns.

Isoquinoline Alkaloids

In 1970 two groups of investigators reported the formation of an isoquinoline alkaloid as a consequence of alcohol metabolism in vitro.[90,91] The discovery was considered to be important in that it demonstrated that mammalian tissues could synthesize alkaloids, in particular the isoquinoline alkaloid, tetrahydropapaveroline, an established intermediate in the biosynthesis of morphine in the poppy plant. The finding, if duplicated in vivo, would provide a conceptual link between alcohol and opiate addiction. If alcohol, during the course of its metabolism, produces a substance that has opioid activity, then it would seem that the ingestion of alcohol, in some individuals, would have a similar effect as the administration of opiates.

Theoretically, tetrahydropapaveroline (THP) could be formed in brain tissue by the condensation of two molecules of the neurotransmitter dopamine. Other isoquinolines could be formed by biogenic amines condensing with molecules of acetaldehyde produced during the metabolism of alcohol. For example, salsolinol could be formed when dopamine condenses with acetaldehyde, and 3-carboxysalsolinol could be formed when DOPA condenses with acetaldehyde. Currently, in vivo and in vitro studies have shown that THP can be biosynthesized,[92] and there is evidence for the formation of isoquinoline alkaloids in some human alcoholics and normals.[93] Sandler et al [94] found salsolinol and THP in the urine of Parkinson patients receiving L-dopa, and THP has been detected in the brains of L-dopa-treated rats that were exposed to ethanol.[95] Recently, Hamilton et al[96] demonstrated the formation of 6-methoxysalsolinol in mice chronically exposed to ethanol. Still other isoquinolines have been demonstrated in pharmacologically manipulated rats during chronic ethanol administration.[97] Thus, there is considerable evidence that isoquinoline biosynthesis occurs in the brain during the administration of alcohol.

The hypothesis that the isoquinolines may provide a link between alcohol and narcotic addiction also led to a number of studies in which attempts were made to demonstrate that the isoquinolines, opiates, and alcohol had common actions; that they enhanced each other's actions; and that their actions were increased or decreased by the same potentiators or antagonists. For example, Marshall et al[98] found that the isoquinoline, 3-carboxysalsolinol, could produce analgesia, potentiate morphine analgesia, and be blocked by the opiate antagonist naloxone. Blum et al,[99] in an extensive review of the literature, accumulated evidence to demonstrate that other isoquinolines demonstrated similar opiate activity. Salsolinol and 3-carboxysalsolinol were also found to enhance the duration of ethanol-induced narcosis in mice.[100] The introduction of the isoquinoline, tetrahydropapaveroline (THP), directly into the brains of rats was found to induce a remarkable increase in voluntary ethanol intake.[101,102] This preference for alcohol persisted for as long as six months after THP infusions were stopped. When alcohol-dependent animals were withdrawn from alcohol, the withdrawal symptoms were shown to be ameliorated by morphine[103] and also by the isoquinoline salsolinol.[99] Naloxone was found to inhibit the development of dependence on ethanol,[104] and to block the excitatory effects of ethanol in rats.[105] Naloxone was also found to reduce some of the effects of acute alcohol intoxication in humans.[106] Ross[107] found that ethanol, opiates, and salsolinol act in a similar fashion to deplete calcium in discrete regions of rat brain, an action that could be blocked by naloxone. Using this system, he also demonstrated the occurrence of biochemical cross-tolerance between morphine and ethanol. Some of the actions of the isoquinolines, then, enhance those of alcohol and the opiates, some are shared with those of alcohol and the opiates, and some are blocked by the same antagonists. Furthermore, Cohen[108] reported that the isoquinolines modify the synaptic properties of catecholamines and inhibit MAO activity so that they may be capable of evoking alterations in behavior.

Thus, it may be hypothesized that alcohol or acetaldehyde in combination with dopamine or other biogenic amines may lead to the formation of isoquinoline alkaloids in humans, especially in those humans who become addicted to alcohol. The formation of the isoquinolines may be enhanced by increased alcohol intake, by increased levels of the biogenic amines, or by both of these. Once the isoquinolines are formed, they could produce long-term, continued preference for alcohol intake. This hypothesis is quite compatible with the previous psychological and biological data that have been presented. For example, one would expect that in individuals with reduced levels of monoamine oxidase activity, there would be an accumulation of biogenic amines and a greater production of isoquinoline alkaloids when ethanol is ingested. The findings that the effects of alcohol are influenced by the opiates and their antagonists also seem to suggest the hypothesis that the opiates and alcohol may ultimately share a common receptor through intermediates, such as the isoquinolines and/or the endorphins.

Endorphins

To examine the possible role of endorphins in the action(s) of alcohol, the endorphins will be considered pharmacologically equivalent to exogenous opiate agonists, such as morphine, heroin, or methadone. This pharmacological quasi-equality, which is experimentally substantiated,[109-112] provides a rich background of research data on opiate agonists. An examination of the pharmacology and neurochemistry of the opiates may help to define the physiological role of the endorphins which, in turn, may suggest relationships to both opiate and alcohol abuse.

The effects of alcohol have often been compared to those of the opiates. Both of these substances have the ability to cause analgesia, euphoria, and tranquilization,[113,114] and both have other similar neurochemical and physiological effects.[115] For example, Wikler et al[116] reported that a 60-ml dose of 95% ethanol in man raises the pain threshold approximately 35% to 40% while not altering other sensory perceptions. Its euphorigenic effect possibly might be the principal property of ethanol that causes dependence liability. The euphoria is a "high" and is manifested by a diminution of anxiety, which results in both a positively reinforcing relaxed and a physically active state.

The paradoxical question of whether alcohol is a stimulant or a depressant has been long debated. According to a respected textbook of pharmacology, alcohol is not a stimulant but, like other general anesthetics, it is a primary and continuous depressant of the central nervous system.[113] The mechanism of action for the stimulating effect of alcohol often is explained by a depressant effect. The theory states that the activity of certain inhibitory centers in the CNS is depressed by low doses of alcohol, which results in a "false sense" of stimulation.[113] This simple explanation, although logical and possibly partially true, has certain shortcomings. First, it presumes that inhibitory neurons are more sensitive to alcohol than other neurons. Second, it is not clear how the inhibition of the inhibitory centers, in addition to stimulation, also causes analgesia and anti-anxiety effects. An alternative explanation of the euphorigenic effects of alcohol would be provided if alcohol could be shown to release endogenous opiate-like substances that could produce the known opiate pharmacological effects, ie, analgesia and anxiety relief.[114] Perhaps the stimulatory effects of alcohol could then be viewed as a counteraction to previous anxiety-related immobilization. At present, direct evidence to support this theory is not at hand. However, we have presented a considerable body of indirect evidence suggesting that the isoquinolines may serve to link alcohol and opiate effects. Furthermore, interesting and provocative reports in the literature support the possibility of an interaction between alcohol and the endorphins.

Lorens and Sainati[105] found that naloxone, a pure opiate antagonist, injected one hour after ethanol administration immediately reversed the excitatory effects of ethanol in the rat. They reasoned that ethanol might release an endogenous opioid whose action at opiate receptors would provide a rewarding effect similar to that produced by direct stimulation of the centers in the lateral hypothalamus. The implication is that endorphin release might underlie the positively reinforcing property of ethanol. If this hypothesis is tenable, opiate antagonists (naloxone or naltrexone) should be able to block alcohol-induced analgesia, euphoria, and motor incoordination, and possibly improve alcoholic stupor by displacing endorphins from opiate receptor sites. A preliminary study of Schenk et al[106] provided encouraging results.

Four human alcoholic subjects with blood alcohol levels ranging from 150 to 320 mg% were selected for study. Following the administration of naloxone at various doses (4 to 28 mg), the authors observed that the nearly unconscious alcoholic subjects spontaneously awoke and were able to talk and perform adequately on motoric tests and coordination tasks. Previously unresponsive corneal reflexes could again be elicited. Changes in blood gas analysis were also striking: pO_2 levels, 15 minutes after naloxone, increased from 58 to 104 mm Hg, while pCO_2 levels dropped from 61 to 48 mm Hg. In comparison, it is interesting to note that β-endorphin has been shown to produce respiratory depression in dogs, which effect was also promptly reversible by naloxone administration as manifested by pO_2 increase and pCO_2 decrease.[117] The question generated by these studies is whether the isoquinolines or the endorphins are being displaced by naloxone in the process of reversing the alcohol-related effects. Certainly, the presence and biosynthesis of endorphins in mammalian brains under normal physiological states is supported by solid experimental data,[118] while similar evidence for isoquinolines is still not as convincing.

If ethanol's reinforcing effects are exerted through the release of endorphins, it is not surprising that numerous investigators have found similarities in the actions of opiates and alcohol.[119,115] The similarity in their reinforcing properties, for example, may explain why animals are willing to substitute opiates for alcohol[120] and alcohol for opiates.[121] Theoretically, alcohol may be viewed as having an endorphin-releasing action, in addition to its dose-dependent and continuous CNS depressive action. Opiates, too, may release endorphins, but they also have the ability to bind and activate the opiate receptors directly. Thus, the reinforcing property of alcohol would be dependent on endorphins, whereas this would not necessarily be the case for opiates. In chronic alcoholism, then, endorphins would be expected to be depleted (depletion tolerance) resulting in the loss of euphorigenic effects. In chronic opiate dependence, occupation of the opiate receptors by exogenous opioids would cause desensitization of the receptors (pharmacological tolerance) and their reinforcing effects would also be lost unless the dose were increased. Although the mechanisms of action are different, chronic excessive use of both alcohol and opiates could result in attenuation or abolition of stimulation of reward centers in the limbic system and decrease the effects for which both drugs were initially consumed. The reinforcing effects of drugs may be preserved by moderate and episodic use. For example, the literature indicates that numerous opiate abusers who were "chipping" or taking opiates sporadically did not develop tolerance nor did they develop the "street addict" personality.[122,123] The parallel character in the alcohol scene is the light-to-moderate social drinker. Problems of opiate and alcohol abuse are recognized only when the habit becomes overwhelming and interferes with the individual's daily functioning.

The question is why these substances are used or, from the legal point of view, abused by humans. If both alcohol and opiates act through the opiate receptors, the answer is not that difficult. Opiates and endorphins are able to divert attention from emotional as well as physical pain. This ability of the opiates is responsible for their excellent analgesic properties and is highly valued for the treatment of intractable pain.[114] Common stress, related to the responsibilities of daily living, is often interpreted as emotionally painful. Some people tolerate this stress reasonably well, some have varying degrees of difficulty, and some may even have major breakdowns.[124] These differences in individual stress tolerance may in the future be found to be related to endorphin physiology; at present, however, the involvement of endorphins, based on opiate agonist behavioral effects, remains speculative.[125,126] The use of alcohol or opiates to escape intolerable emotional stress may represent an attempt to attain psychological homeostasis in the CNS through the physiological effect of the endorphins.[127] Endorphins may be released from storage vesicles in response to major stressful experiences, or they may be chemically released, possibly by taking in alcohol, to provide a defense mechanism and reduce emotional pain during crises. If this role for the endorphins can gain experimental support, mental illness characterized as intolerance of stress may well come to be classified as another metabolic disorder similar in mechanism to diabetes. In diabetes, exogenous insulin is administered for insulin deficiency; in the treatment of psychosis, an exogenous opiate may become useful for endorphin deficiency.[127]

The choice of the type of drug used by a given individual is influenced by external and internal forces. External forces include cultural background and social mores. For example, Oriental cultures historically have been permissive with respect to the use of opium, which promotes passive adaptation of the individual to his environment. In comparison, Western cultures have been permissive regarding the use of alcohol, which, among other effects, tends to reduce inhibition and release aggressive behavior. In Oriental cultures the use of alcohol and in Western cultures the use of opiates are considered to be flagrant abuses.[126] Nevertheless, the fact that users of the nonapproved drugs do exist in both cultures directs our attention toward a consideration of internal as well as external forces. It has been suggested that individuals will use drugs that help them feel better.[128] According to Pittel,[129] the user's psychological constitution is a major factor in the choice of drugs. Dependence on either alcohol or opiates is likely to occur in individuals who lack the psychological stability to cope with inner conflicts and environmental frustrations. In such individuals the drugs tend to compensate for the absence of inner strength and structure. Of all drugs, the opiates seem to have the strongest reinforcing effect.[130] Presumably this effect occurs through the stimulation of the reward centers in the limbic system. If the endorphins act in the same manner as opiates, it may be shown in the future that the effects of alcohol and perhaps other euphoria-producing drugs are achieved through endorphin release in the CNS. Thus, although opiate and alcohol dependence superficially appear different, both are primary reinforcers and both may act by stimulating, either directly or indirectly, the same opiate receptors.

DISCUSSION

Over the years, there have been changing fashions with respect to whether the similarities or the differences between alcohol and drug abuse have been emphasized. For example, alcohol and opium have been seen as having similar and beneficial social, medicinal, and mind-altering effects when used in moderation. Both have also been seen as having similar disastrous consequences when used in excess. Sharing many of the same characteristics, including the tendency to induce physical dependence and tolerance, alcohol and narcotic abuse have, nevertheless, often been viewed as distinct and different entities. Few "drunkards" were expected to resort to drugs, and few "dope fiends" to be satisfied by alcohol. Narcotic addiction was considered Oriental and alcoholism Occidental.

At the turn of the twentieth century the psychoanalysts, searching for broad psychodynamic explanations of behavior, were impressed by the many similarities they noted in the personal histories of alcohol and narcotic abusers. Drinking, drug taking, smoking, overeating, and even gambling were conceptualized as related impulse disorders deriving from difficulties encountered during the oral stage of psychosexual development. Since the impulse disorders were considered to be resistant to treatment, psychiatrists, for the most part, showed little interest in them.

In recent decades, however, under the impetus of the National Council on Alcoholism and the drug abuse explosion of the 1960s, alcoholism and drug addiction seemed to emerge as new fields. Representing separate constituencies, it was not surprising that the National Institute on Alcoholism and Alcohol Abuse (NIAAA) and the National Institute on Drug Abuse (NIDA) developed educational, research, prevention, and treatment programs in parallel, rather than integrated with those of the National Institute of Mental Health (NIMH). In contrast to the psychodynamic approach, which had emphasized *similarities among individuals* addicted to different substances, programs supported by separate institutes led to a greater focus on *differences between the substances* to which individuals were addicted. As a consequence, despite some notable exceptions, there appears to have been little overlap between the fields of alcoholism and drug addiction with respect to educational programs, educators, training activities, support constituencies, funding streams, treatment personnel, treatment systems, prevention programs, investigators, journals, and research ideas and studies. However, there are some indications that we may be on the verge of still another shift in focus. A number of psychological and biochemical studies are again pointing to common aspects of the addictions. We are recognizing an ever increasing number of polydrug abusers who, instead of becoming addicted to a particular substance, take a wide variety of drugs including alcohol. In addition, the Eagleville Hospital Rehabilitation Center has recently pioneered the combined treatment of addicted individuals and demonstrated that alcohol and drug abusers could be effectively treated together.[131]

Our purpose here, then, was to review some of the findings and studies that are reemphasizing the similarities between alcoholism and drug addiction. As such, it should be made explicit that this review is neither exhaustive nor unbiased.

Surveys of adolescent, male and female, alcohol and drug users and abusers indicated that they came from similar deprived and frustrating backgrounds and exhibited similar maladaptive characteristics and behaviors. They were difficult to differentiate from one another and they shared many sociopathic and depressive features. It was found that those who used more alcohol also used more drugs. Several of the studies demonstrated that certain personality traits and family relationship patterns existed and could be recognized prior to the development of problem drinking and alcoholism. These characteristics could be used, then, to identify vulnerable individuals early and to plan intervention strategies aimed at secondary prevention.

Some of the traits that were typical of the young alcohol and drug abusers, such as impulsiveness, restlessness, and aggressiveness, were reminiscent of the symptoms of individuals with minimal brain dysfunction. It has been suggested, on the basis of some interesting studies of alcoholic patients, that certain individuals may inherit a neuropsychological deficit which makes them vulnerable to the development of MBD and to alcoholism. This work has not yet been extended to drug-dependent patients. If a biochemical mechanism were found to underlie the hypothesized deficit, then one might expect that there are groups of patients who "self-medicate" with alcohol or opiates to compensate for this underlying neurobiological defect. It is certainly not uncommon to observe patients who are clinically "normalized" by the administration of alcohol or opiates.

Psychodynamic and psychometric descriptions of adult alcoholics and drug addicts were very similar to each other and to the descriptions of adolescent alcohol and drug users and abusers. Again, features of sociopathy and affective illness were found to be prominent in the profiles of these patients. Relationships between the Zuckerman Sensation Seeking Scale and alcohol use, drug use, hypomania, sociopathy, and decreased platelet monoamine oxidase activity further suggested the interesting possibility of linkages between the psychology and biology of these conditions.

There is considerable evidence for the genetic transmission of alcoholism, affective illness, and sociopathy, and some studies indicate overlapping vulnerabilities among the three conditions. While no clear evidence is reported for the genetic transmission of opiate dependence, a similar vulnerability is suggested because of the many shared characteristics between drug addiction and alcoholism, affective illness, and sociopathy. In addition, some evidence suggests that the Sensation Seeking Scale, which relates to all four conditions, may have a genetic basis. Finally, MAO activity, which plays a crucial role in neurotransmitter activity, may be involved in the biological mechanism that is inherited and is responsible for the common behavioral manifestations that have been noted in these conditions. If so, one could predict that MAO activity would be found to be low in sociopathic and opiate-dependent individuals.

The possibility that many of the actions of alcohol and opiates are mediated by a common biological mechanism was reinforced by the recent findings that isoquinoline alkaloids could be formed in brain tissue by the condensation of biogenic amines with acetaldehyde produced during the metabolism of alcohol, and that at least one of these isoquinolines was a known intermediate in the biosynthesis of morphine. The isoquinolines were found to share some of the actions of alcohol and the opiates and to enhance others, while some of their actions were blocked by the same antagonists. It was also demonstrated that very minute amounts of the isoquinolines could induce a continued and long-term preference for alcohol intake. In individuals with reduced levels of monoamine oxidase, then, there could be an accumulation of dopamine and other biogenic amines and a greater production of the isoquinolines when alcohol is ingested. In turn, the increase in isoquinolines could lead to an increased preference for alcohol.

A considerable body of indirect evidence has been collected suggesting that alcohol may act to stimulate the release of endorphins, which could also account for the shared effects of alcohol and the opiates and possibly the isoquinolines as well. For example, naloxone was found to rapidly reverse the excitatory effects of ethanol and inhibit the development of alcohol dependence in several animal studies, and to reduce the effects of alcohol intoxication in an interesting human study. It was also shown to block a similar calcium-depleting effect of ethanol, salsolinol, and the opiates and to reverse a respiratory depressant action of ethanol and β-endorphin. Such findings have led a number of investigators to believe that alcohol, through the release of endorphins, may act on opiate receptors, whereas the opiates may act directly or through the release of endorphins. The reinforcing properties of alcohol and the opiates may be due then to their direct or indirect actions on common receptors in the various reward centers of the limbic system.

Our review began with observations of similarities in the backgrounds and behavior of alcohol and other drug abusers, moved on to studies suggesting the possibility of a common, inherited biological deficit, and ended with findings which seemed to implicate monoamine oxidase, the isoquinolines, and the endorphins as intermediaries acting on the neurotransmitters and on common reward center receptors. The use of alcohol and opiates could then be seen as attempts at "self-medication" to compensate for a biological deficit and reduce other psychiatric disturbances.

As we assembled the materials for this chapter comparing alcohol and opiate addiction, we became aware of many informational gaps and many interesting research questions. For example, we do not know whether monoamine oxidase activity levels are low in drug addicts and young delinquents. We found almost no genetic studies of drug addiction. Possibly because most drug addicts in the 1960s came from families in which there was no previous drug addication, such studies were felt to be unnecessary. But what if drug addiction is genetically linked to alcoholism? It may be that the drug abusers did not really represent new cases but merely the use of a new substance. Was there a high incidence of affective illness or sociopathy in their families? These are researchable questions, and there are many more. We did not even consider comparisons with amphetamine and barbiturate dependence.

Despite the similarities that have been noted between alcohol and opiate addiction, we are not proposing that they are manifestations of a single disorder. Indeed, there are many dissimilarities between them, ranging from differences in tissue toxicity and symptoms of intoxication to racial preferences. We are more concerned that researchers do not consider them in isolation but focus, instead, on more clearly delineating their similarities and differences. Many of the studies that were cited have not been replicated, many of the linkages are hypothetical, and the hypotheses are rapidly changing and becoming more sophisticated in this burgeoning field. As the data continue to accumulate, however, there is increasing hope that we shall become better able to appreciate genetic patterns, identify vulnerable youngsters, develop prevention strategies, differentiate addicted subgroups, and more appropriately select treatments for particular patient needs and problems.

REFERENCES

1. McCord W, McCord J, Gudeman J: *Origins of Alcoholism.* Stanford, Calif, Stanford University Press, 1960.

2. DeLint J: Alcoholism, birth rank and parental deprivation. *Am J Psychiatry* 120:1062–1065, 1964.

3. Lindbeck VL: The woman alcoholic—A review of the literature. *Int J Addict* 7:567–580, 1972.
4. Demone HE: Implications from research on adolescent drinking, in *1961 Alcohol Education Conference Proceedings.* Washington, DC, Dept of HEW, 1966.
5. Jessor R, Jessor SL: Problem drinking in the young: Personality, social and behavioral antecedents and correlates, in *Proceedings Second Annual Alcoholism Conference of the National Institute of Alcohol Abuse and Alcoholism.* DHEW Pub #(NIH) 74-676, pp 3-23, 1973.
6. Rotter JB: *Social Learning and Clinical Psychology.* Englewood Cliffs, NJ, Prentice-Hall, 1954.
7. Rotter JB, Chance JE, Phares EJ: *Applications of a Social Learning Theory of Personality.* New York, Holt, Rinehart and Winston, 1972.
8. Hamburg BA, Kraemer HC, Jahnke W: A hierarchy of drug use in adolescence: Behavioral and attitudinal correlates of substantial drug use. *Am J Psychiatry* 132:1155–1163, 1975.
9. Ausubel DP: Causes and types of narcotic addiction: A psychosocial view. *Psychiatry Q* 35:523–531, 1961.
10. Gilbert JG, Lombardi DN: Personality characteristics of young male narcotic addicts. *J Consult Psychol* 31:536–558, 1967.
11. Glasscote R, Sussex JN, Jaffe JH, et al: *The Treatment of Drug Abuse.* Washington, DC, Joint Information Service of the American Psychiatric Association, 1972.
12. Chinlund S: The female addict. *Science News* 95:578, 1969.
13. Wechsler H, Thum D: Alcohol and drug abuse among teenagers—A questionnaire study, in *Proceedings Second Annual Alcoholism Conference of the National Institute on Alcohol Abuse and Alcoholism.* DHEW Pub #(NIH) 74-676, 1973.
14. Westermeyer J, Walzer V: Sociopathy and drug abuse in a young psychiatric population. *Dis Nerv Sys* 36:673–677, 1975.
15. Renshaw D: *The Hyperactive Child.* Chicago, Hall, 1974.
16. Mendelson W, Johnson N, Stewart MA: Hyperactive children as teenagers: A follow-up study. *J Nerv Ment Dis* 153:273–279, 1971.
17. Morrison JR, Stewart MA: A family study of the hyperactive child syndrome. *Biol Psychiatry* 3:189–195, 1971.
18. Morrison JR, Stewart MA: The psychiatric status of the legal families of adopted hyperactive children. *Arch Gen Psychiatry* 28:888–891, 1973.
19. Morrison JR, Stewart MA: Evidence for polygenetic inheritance in the hyperactive child syndrome. *Am J Psychiatry* 130:791–792, 1973.
20. Goodwin DW, Schulsinger LF, Hermansen L, et al: Alcoholism and the hyperactive child syndrome. *J Nerv Ment Dis* 160:349–353, 1975.
21. Tarter RE: Empirical investigations of psychological deficit, in Tarter RE, Sugarman AA (eds): *Alcoholism: Interdisciplinary Approaches to an Enduring Problem.* Reading, Mass, Addison-Wesley, 1976.
22. Tarter RE, McBride M, Buopane N, et al: Differentiation of alcoholics according to childhood history of minimal brain dysfunction, family history and drinking pattern. *Arch Gen Psychiatry* 34:761–768, 1977.
23. Seldon NE: The family of the addict: A review of the literature. *Int J Addict* 7:97–107, 1972.
24. Knight RP: Psychodynamics of chronic alcoholism. *J Nerv Ment Dis* 86:538–548, 1937.
25. Menninger KA: *Man Against Himself.* New York, Harcourt Brace, 1938.

26. Fenichel O: *The Psychoanalytic Theory of Neuroses.* New York, WW Norton, 1945.

27. Rado S: Psychoanalysis of pharmacothymia. *Psychoanal Q* 2:1–23, 1933.

28. Knight RP, Prout C: A study of results in hospital treatment of drug addiction. *Am J Psychiatry* 108:303–308, 1951.

29. Brill L: Some notes on dynamics and treatment in narcotic addiction. *J Psych Soc Work* 23:67–81, 1954.

30. Hartman D: A study of drug taking adolescents. *Psychoanal Study Child* 24:384–430, 1969.

31. Sharoff R: Character problems and their relationship to drug abuse. *Am J Psychoanal* 29:286–293, 1969.

32. Reith G, Crockett D, Craig K: Personality characteristics in heroin addicts and non-addicted prisoners using the Edwards personality preference schedule. *Int J Addict* 10:97–112, 1975.

33. Hill H, Hartzen C, Davis H: An MMPI factor analytic study of alcoholics, narcotic addicts and criminals. *Q J Stud Alcohol* 23:411–431, 1962.

34. Heller ME, Mordkoff AM: Personality attributes of the young non-addicted drug abuser. *Int J Addict* 7:65–72, 1972.

35. Olson RW: MMPI sex differences in narcotic addicts. *J Gen Psychol* 71:157–266, 1964.

36. Overall JE: MMPI personality patterns of alcoholics and narcotic addicts. *Q J Stud Alcohol* 34:104–111, 1973.

37. Lanyon RI: *A Handbook for MMPI Group Profiles.* Minneapolis, University of Minnesota Press, 1968.

38. Gilberstadt H, Duker J: *A Handbook of Clinical and Actuarial MMPI Interpretation.* Philadelphia, Saunders, 1965.

39. Goldstein SG, Linden JD: Multivariate classification of alcoholics by means of the MMPI. *J Abnorm Psychol* 74:661–669, 1969.

40. Weissman MM, Slobetz F, Prusoff B, et al: Clinical depression among narcotic addicts maintained on methadone in the community. *Am J Psychiatry* 133:1434–1438, 1976.

41. Senay E: *Depression in Drug Abusers.* Presentation at annual meeting of the American Psychiatric Association, Toronto, Canada, 1977.

42. Apfeldorf M: Alcoholism scales of the MMPI: Contributions and future directions., *Int J Addict* 13:17–53, 1978.

43. MacAndrew C: The differentiation of male alcoholic outpatients from non-alcoholic psychiatric patients by means of the MMPI. *Q J Stud Alcohol* 26:238–246, 1965.

44. Kranitz L: Alcoholics, heroin addicts and non-addicts: Comparison on the MacAndrew alcoholism scale of the MMPI. *Q J Stud Alcohol* 33:807–809, 1972.

45. Herl D: A study of select cognitive orientations among a drug-abuse, non-abuse, and normal adolescent population. *Ariz Med* 29:408–412, 1972.

46. Zuckerman M: Drug usage as one manifestation of a "sensation seeking" trait, in Keup W (ed): *Drug Abuse: Current Concepts and Research.* Springfield, Ill, Charles C Thomas, 1972.

47. Zuckerman M: The sensation seeking motive, in Maher BA (ed): *Progress in Experimental Personality Research* vol 7. New York, Academic Press, 1974.

48. Zuckerman M: Sensation seeking and psychopathy, in Hare RD, Schalbring D (eds): *Psychopathic Behaviour: Approaches to Research.* New York, John Wiley & Sons, 1978.

49. Murphy DL, Weiss R: Reduced monoamine oxidase activity in blood platelets from bipolar depressed patients. *Am J Psychiatry* 128:351–357, 1972.

50. Schooker C, Zahn TP, Murphy DL, et al: Psychological correlates of monoamine oxidase in normals. *J Nerv Ment Dis* 166:177–186, 1978.

51. Rosenthal D: *Genetic Theory and Abnormal Behavior.* New York, McGraw-Hill, 1970.

52. Goodwin DW: Adoption studies of alcoholism. *J Operational Psychiatry* 7:54–63, 1976.

53. Cotton NS: The familial incidence of alcoholism: A review. *J Stud Alcohol* 40:89–116, 1979.

54. Kaij L, Dock J: Grandsons of alcoholics: A test of sex-linked transmission of alcohol abuse. *Arch Gen Psychiatry* 32:1379–1381, 1975.

55. Kaij L: *Alcoholism in Twins: Studies on the Etiology and Sequels of Abuse of Alcohol.* Stockholm, Almquist and Wiksell, 1960.

56. Goodwin DW, Schulsinger LF, Hermansen L, et al: Alcohol problems in adoptees raised apart from alcoholic biological parents. *Arch Gen Psychiatry* 23:238–243, 1973.

57. Goodwin DW, Schulsinger LF, Moller N, et al: Drinking problems in adopted and non-adopted sons of alcoholics. *Arch Gen Psychiatry* 31:164–169, 1974.

58. Bohman M: Genetic aspects of alcoholism and criminality. *Arch Gen Psychiatry* 35:269–276, 1978.

59. Goodwin DW, Schulsinger LF, Knop J, et al: Alcoholism and depression in adopted-out daughters of alcoholics. *Arch Gen Psychiatry* 34:751–755, 1977.

60. Feighner JP, Robins E, Guze SB, et al: Diagnostic criteria for use in psychiatric research. *Arch Gen Psychiatry* 29:57–63, 1972.

61. Schuckit M, Goodwin DW, Winokur G: A study of alcoholism in half-siblings. *Am J Psychiatry* 128:1132–1136, 1972.

62. Winokur G, Reich T, Rimmer J, et al: Alcoholism III. Diagnosis and familial psychiatric illness in 259 alcoholic probands. *Arch Gen Psychiatry* 27:104–111, 1970.

63. Pottenger M, McKernon J, Patrie LE, et al: The frequency and persistence of depressive symptoms in the alcohol abuser. *J Nerv Ment Dis* 166:562–570, 1978.

64. Jellnick EM: *The Disease Concept of Alcoholism.* Highland Park, ND, Hillhouse Press, 1960.

65. Sundby P: *Alcoholism and Mortality.* National Institute for Alcohol Research, Pub 6, Universitetsforlaget, Oslo, 1967.

66. Guze SB, Robins E: Suicide and primary affective disorders. *Br J Psychiatry* 117:437–438, 1970.

67. Maddox JF, Elliott B: Problem drinking among patients on methadone. *Am J Drug Alcohol Abuse* 2:245–254, 1975.

68. Rosanoff AJ, Handy L, Plasset IR: The etiology of manic depressive syndromes with special reference to their occurrence in twins. *Am J Psychiatry* 91:725–762, 1935.

69. Kallman FJ: Genetic principles in manic-depressive psychosis, in Zubin J (ed): *Depression.* New York, Grune and Stratton, 1954.

70. Allen MG, Cohen S, Pollin W, et al: Affective illness in veteran twins: A diagnostic review. *Am J Psychiatry* 131:1234–1239, 1974.

71. Reich T, Clayton PJ, Winokur G: Family history studies: V. The genetics of mania. *Am J Psychiatry* 123:1358–1369, 1969.

72. Dunner DL, Gershon ES, Goodwin FK: Heritable factors in the severity of affective illness. *Sci Proc Am Psych Assoc* 123:187–188, 1970.

73. Gershon ES, Mark A, Cohen N, et al: Transmitted factors in the morbid risk of affective disorders. *J Psychiatr Res* 12:283–299, 1975.

74. James N, Chapman CJ: A genetic study of bipolar affective disorders. *Br J Psychiatry* 126:449–456, 1975.

75. Yoshimasu S: Criminal life curves of monozygotic twin-pairs. *Acta Criminologiae et Medicae Legalis Japonica* 31:5–6, 1965.

76. Christianson KC: Crime in a Danish twin population. *Acta Geneticae Medicae et Gamellologiae,* 19:323–326, 1970.

77. Crowe RR: An adoption study of antisocial personality. *Arch Gen Psychiatry* 31:785–791, 1974.

78. Crowe RR: An adoptive study of psychopathy: Preliminary results from arrest records and psychiatric hospital records, in Fieve RR, Rosenthal D, (eds): *Genetic Research in Psychiatry.* Baltimore, Johns Hopkins University Press, 1975.

79. Cadoret RJ: Psychopathology in adopted away offspring of biologic parents with antisocial behavior. *Arch Gen Psychiatry* 35:176–184, 1978.

80. Cloninger CR, Reich T, Guze SB: The multifactorial model of disease transmission: II. Sex differences in the family transmission of sociopathy (antisocial personality). *Br J Psychiatry* 127:11–22, 1975.

81. Stahl SM: The human platelet: A diagnostic and research tool for the study of biogenic amines in psychiatric and neurological disorders. *Arch Gen Psychiatry* 34:509–516, 1977.

82. Leckman JF, Gershon ES, Nichols AS, et al: Reduced MAO activity in first-degree relatives of individuals with bipolar affective disorders. *Arch Gen Psychiatry* 34:601–606, 1977.

83. Takahashi S, Tani N, Yamane H: Monoamine oxidase activity in blood platelets in alcoholism. *Folia Psychiatr Neurol* 30:453–462, 1977.

84. Weiberg A, Gottfries CG, Oreland L: Low platelet monoamine oxidase activity in human platelets. *Med Biol* 55:181–186, 1977.

85. Sullivan JL, Stanfield CN, Schanberg S, et al: Platelet monoamine oxidase and serum dopamine-B-hydroxylase activity in chronic alcoholics. *Arch Gen Psychiatry* 34:409–516, 1978.

86. Gottfries CG, Oreland L, Weiberg A, et al: Lowered monoamine oxidase activity in brains from alcoholic suicides. *J Neurochem* 25:667–673, 1975.

87. Grote SS, Moses SG, Robins E, et al: A study of selected catecholamine metabolizing enzymes: A comparison of depressive suicides and alcoholic suicides with controls. *J Neurochem* 23:791–802, 1974.

88. Wyatt RJ, Murphy DL, Belmaker R, et al: Reduced monoamine oxidase in platelets: A possible genetic marker for vulnerability to schizophrenia. *Science* 179:916–918, 1973.

89. Nies A, Robinson DS, Harris LS, et al: Comparison of monoamine oxidase substrate activities in twins, schizophrenics, depressives and controls. *Adv Biochem Psychopharmacol* 12:59–70, 1974.

90. Cohen G, Collins M: Alkaloids from catecholamines in adrenal tissue: Possible role in alcoholism. *Science* 167:1749–1751, 1970.

91. Davis VE, Walsh MJ: Alcohol, amines and alkaloids: A possible basis for alcohol addiction. *Science* 167:1005–1007, 1970.

92. Weiner H: Relationship between 3, 4-dihydroxyphenylacetaldehyde levels and tetrahydropapaveroline formation. *Alcohol Clin Exp Res* 2:127–131, 1978.

93. Nijm WP, Riggin R, Teas G, et al: Urinary dopamine-related tetrahydroisoquinolines: Studies of alcoholics and non-alcoholics. *Fed Proc* 36:343, 1977.

94. Sandler M, Carter SB, Hunter KR: Tetrahydroisoquinoline alkaloids: In vivo metabolites in men. *Nature* 241:439–443, 1973.

95. Turner AJ, Baker KM, Algeri S, et al: Tetrahydropapaveroline: Formation in vivo and in vitro in rat brain. *Life Sci* 14:2247–2257, 1974.

96. Hamilton MG, Blum K, Hirst M: Identification of an isoquinoline alkaloid after chronic exposure to ethanol. *Alcohol Clin Exp Res* 2:133–137, 1978.

97. Collins MA, Bigdeli MG: Tetrahydroisoquinolines in vivo: I. Rat brain formation of salsolinol, a condensation product of dopamine and acetaldehyde, under certain conditions during ethanol intoxication. *Life Sci* 16:585–602, 1976.

98. Marshall A, Hirst M, Blum K: Morphine analgesia augmentation by and direct analgesia with e-carboxysalsolinol. *Experientia* 33:754–755, 1977.

99. Blum K, Hamilton MG, Hirst M, et al: Putative role of isoquinoline alkaloids in alcoholism. A link to opiates. *Alcohol Clin Exp Res* 2:113–210, 1978.

100. Marshall A, Hirst M: Potentiation of ethanol narcosis by dopamine-L-dopa-based isoquinolines. *Experientia* 32:201–203, 1976.

101. Meyers RD, Oblinger MM: Alcohol drinking in the rat induced by acute intra-cerebral infusion of two tetrahydroisoquinolines and a beta-carboline. *Drug Alcohol Depend* 2(5–6):469–483, 1977.

102. Meyers RD, Melchior CL: Alcohol drinking; abnormal intake caused by tetrahydropapaveroline in brain. *Science* 196:554–556, 1977.

103. Blum K, Wallace JE, Schwertner HA, et al: Morphine suppression of ethanol withdrawal in mice. *Experientia* 32:79–82, 1976.

104. Blum K, Wallace JE, Futterman SL, et al: Naloxone-induced inhibition of ethanol dependence in mice. *Nature* 265:49–51, 1977.

105. Lorens SA, Sainati SM: Naloxone blocks the excitatory effect of ethanol and chlordiazepoxide on lateral hypothalamic self-stimulation behavior. *Life Sci* 23:1359–1364, 1978.

106. Schenk GK, Enders P, Engelmeier MP, et al: Application of the morphine antagonist naloxone in psychiatric disorders. *Arzneim Forsch* 28:1274–1277, 1978.

107. Ross DH: Inhibition of high affinity calcium binding by salsolinol. *Alcohol Clin Exp Res* 2:139–144, 1978.

108. Cohen G: The synaptic properties of some tetrahydroisoquinoline alkaloids. *Alcohol Clin Exp Res* 2:121–126, 1978.

109. Loh HH, Tseng LF, Wei E, et al: β-endorphin is a potent analgesic agent. *Proc Natl Acad Sci* 73:2895–2898, 1976.

110. Graf L, Szekely JI, Ronai AZ, et al: Comparative study on analgesic effect of met^5-enkephalin and related lipotropin fragments. *Nature* 263:240–242, 1976.

111. Wei E, Loh H: Physical dependence on opiate-like peptides. *Science* 193:1262–1263, 1976.

112. Plotnikoff NP, Kastin AJ, Coy DU, et al: Neuro-pharmacological actions of enkephalin after systemic administration. *Life Sci* 19:1283–1288, 1976.

113. Richie MJ: The aliphatic alcohols, in Goodman LS, Gilman A (eds): *The Pharmacological Basis of Therapeutics.* New York, Macmillan, 1975.

114. Jaffe JH, Martin WR: Narcotic analgesics and antagonists, in Goodman LS, Gilman A (eds): *The Pharmacological Basis of Therapeutics,* ed 5. New York, Macmillan, 1975.

115. Blum K (ed): *Alcohol and Opiates, Neurochemical and Behavioral Mechanisms.* New York, Academic Press, 1977.

116. Wikler A, Godell H, Wolff HG: Studies on pain: The effects of analgesic agents on sensations other than pain. *J Pharmacol Exp Ther* 83:294–299, 1945.

117. Moss IR, Freidman E: β-endorphin: Effects on respiratory regulation. *Life Sci* 23:1271–1276, 1978.

118. Goldstein A: Opioid peptides: Endorphins in pituitary and brain. *Science* 193:1081–1086, 1976.

119. Ross DH, Medina MA, Cardenas HL: Morphine and ethanol: Selective depletion of regional brain clacium. *Science* 186:63–65, 1974.

120. Sinclair JD, Adkins J, Walker S: Morphine-induced suppression of voluntary alcohol drinking in rats. *Nature* 246:425–427, 1973.

121. Smith SG, Werner TE, Davis WM: Intravenous drug self-administration in rats: Substitution of ethyl alcohol for morphine. *Psychol Rec* 25:17–20, 1975.

122. Gay GR, Winker JJ, Newmeyer JA: Emerging trends of heroin abuse in the San Francisco Bay area. *J Psychedelic Drugs* 4:53–55, 1971.

123. Powell DH: A pilot study of occasional heroin users. *Arch Gen Psychiatry* 28:586–594, 1973.

124. Hartman E: Schizophrenia: A theory. *Psychopharmacology* 49:1–15, 1976.

125. Khantzian EJ: Opiate addiction: A critique of theory and some implications for treatment. *Am J Psychother* 28:59–70, 1974.

126. Jaffe JH: Drug addiction and drug abuse, in Goodman LS, Gilman A (eds): *The Pharmacological Basis of Therapeutics,* ed 5. New York, Macmillan, 1975.

127. Verebey K, Volavka J, Clouet DH: Endorphins in psychiatry: An overview and a hypothesis. *Arch Gen Psychiatry* 35:877–888, 1978.

128. Nyswander M: *The Drug Addict as a Patient.* New York, Grune and Stratton, 1956, pp 57–82.

129. Pittel SM: Psychological aspects of heroin and other drug dependence. *J Psychedelic Drugs* 4:40–45, 1971.

130. Wikler A: A psychodynamic study of a patient during experimental self-regulated re-addiction to morphine. *Med Biol* 55:181–186, 1977.

131. Ottenberg DJ, Carpey EL (eds): *Proceedings of the 8th Annual Eagleville Conference on Alcoholism and Drug Addiction: Critical Issues in Hard Times.* June 20, 1975. Eagleville Hospital and Rehabilitation Center, Eagleville, Penn, *Amer J Drug Alcohol Abuse* vol 3, 1976.

9 The Law of Alcohol and Drugs: Some Parallels and Incongruities

Douglas A. Eldridge

The law relating to alcohol and drug use reflects the classic struggle in the scheme of social organization known as government: the tension between the liberty of the individual and the need for societal controls. This tension has existed almost from the beginning of recorded history. The use of psychoactive substances antedates even that.*

In ancient Egypt temperance tracts were resorted to in order to discourage the drinking habits of the population:

> Take not upon thyself to drink a jug of beer. Thou speakest and an unintelligible utterance issueth from thy mouth. If thou fallest down and thy limbs break, there is no one to hold out a hand to thee. Thy

*A Chinese text on pharmacology held by tradition to have been written by the Emperor Shen Nung refers to the use of marijuana in 2737 BC. Seeds from the plant were found in a burial place near Berlin, Germany, and are believed to date from 500 BC. (Edward M. Brecher, Consumers Union, *Licit and Illicit Drugs* 1972 pp 397–398.) Various peoples across the world have invented their own alcoholic beverages, ranging from the wine of the Mediterranean peoples to the beer made from roots by primitive African tribes.

> companions in drink stand up and say: "Away with this sot." And thou art like a little child.[1]

Homer sang of the great difficulty Ulysses had in wresting his men away from the Land of the Lotus Eaters:

> As soon as scouts tasted that honey-sweet fruit, they thought no more of coming back to us with news, but chose rather to stay there with the lotus-eating natives, and chew their lotus, and good-bye to home. I brought them back to the ships by main force, grumbling and complaining, and when I had them there, tied them up and stowed them under the benches. Then I ordered the rest to hurry up and get aboard, for I did not want them to have a taste of lotus and say good-bye to home.[2]

The power of psychoactive substances was recognized by early societies who restricted their use to religious occasions. The Bible in speaking of the "vapors from burnt spices and aromatic gums [which] were considered a pleasurable act of worship" could very well mean the marijuana and incense combination so popular today.[3] The Aztecs used peyote in pre-Columbian religious rituals.[4] The practice persists today in the sacramental use of wine. The Greeks and Romans even devoted a god to this power—Bacchus.

However, the attractiveness of alcohol and other psychoactive substances to all levels of society made the restriction to religious use politically unenforceable in many cultures. Other means of control had to be found. This chapter proposes to take an excursion through the legal methods that have been used in the United States to control these substances. Time and time again the vehicle for the excursion appears to be a pendulum, as the legal means of control swing back and forth from complete proscription of use and possession, with penal sanctions for violators, to more lenient approaches acknowledging certain levels of use and providing treatment to abusers of the substances.

Throughout much of this period the methods for controlling alcohol and drugs have been parallel. The present period manifests a significant disparity since, in the public view, there is a need for stiffer controls over drugs than over alcohol. These parallels and incongruities within the public mind are reflected in the laws of the United States.

A PAGE OF HISTORY

It is an aphorism in law that a page of history is worth a volume of logic. To understand the present divergences in the law's treatment of alcohol and drugs, it would be helpful to examine the place of the psychoactive substances in the recent past.

As in most other societies across the globe, alcohol played an essential part in the life of the early colonies. Recall that Jamaican rum was one leg of the triangle trade. Slaves were brought from Africa to the Caribbean on the first leg. After discharging their human cargoes into New World bondage, the ships filled their holds with island rum and sailed to the colonies. With the cash realized from the spirituous cargoes, the ships' captains were able to purchase colonial manufacturies as well as raw natural resources, such as indigo, flax, tobacco, corn, and timber for shipment back to Europe. The mercantile economy would not have worked were it not for the thirst of the colonists.

Even the Puritan founders of the nation accepted alcohol almost as second nature. It was one of the essentials of their diet. When the *Mayflower's* supply of beer was exhausted, no less than William Bradford, the staunch Puritan governor, was heard to complain that "it was unnatural to expect an Englishman to drink water!"

During the Revolution and thereafter, when dollars were scarce in the frontier settlements, wages were often paid in whiskey. This practice was paralleled in the next century when Chinese laborers building the transcontinental railroad were paid in opium.

The nearly universal acceptance of whiskey as part of the life in the early United States underlies the first serious threat to the internal stability of the new nation. The Whiskey Rebellion of 1794 was the response to a law, the Revenue Act of 1791, providing for an excise tax on whiskey. Taxes on alcoholic beverages had not been well received by the public. The cider tax caused an outcry in England in 1733 that was just short of rebellion. A tax on rum raised a similar civil commotion in Massachusetts in 1754.[5] The taxing of indigenous distillers caused a real rebellion in Pennsylvania with large frontier areas of that state refusing to follow the federal law. Alexander Hamilton, the Secretary of the Treasury, assembled 15,000 militia men from four states as a demonstration of seriousness of the infant nation in enforcing its statutes—even revenue measures. The 15,000-man force was larger than had been gathered for any battle in the Revolution other than Yorktown. It was effective. The governmental units were not fired upon and the frontier recalcitrants agreed to recognize the primacy of the federal law.[6]

Nine years after this demonstration of the importance of alcohol to the American way of life, morphine was isloated from opium by a chemist in Germany. Laudanum had been known to the western world since the sixteenth century.[7] In 1816 Samuel Taylor Coleridge composed his great poem, "Kubla Kahn," in an opium dream. By 1821 Thomas

DeQuincey was extolling its use as a recreational drug in the *Confessions of an English Opium Eater.* Throughout the century, the "eating" of opium by the English working class was of sufficient concern to generate a series of governmental examinations of the practice.[8]

The United States underwent a period of religious revivalism during the 1820s and 1830s. One of its manifestations was the rise of the anti-alcohol legislation. Massachusetts attempted to limit the flow of spirits to individuals by prohibiting sales of fewer than 15 gallons. This strange device for discouragement apparently was not effective and was repealed in 1840, two years after its passage. However, the Prohibition spirit was rising and by 1846 Maine had passed such a law. Several other states followed suit before the Civil War.

The Civil War marked the first major recognition of the dangers of drug addiction in the country. With the intervention of the hypodermic syringe in 1853, morphine became the analgesic of choice for serious pain. Indeed, it was used so often on the battlefield and in field hospitals during the Civil War that morphine addiction became known as "the army disease."

The return to relative domestic tranquility brought a new enthusiasm for control. The South had been punished for its wicked ways, and through Reconstruction it was being taught the proper Yankee way to live. With the abolitionists vindicated, the spirit of moralism sought new challenges. Founded in 1869, the Prohibition Party gained strength up until the imposition of national prohibition. Its high water mark was the election of 1892 when, under the banner of General John Bidwell, it gained 2% of the presidential ballot. Even today the party is alive and has fielded presidential candidates in every election since its formation.

The heightened concern for control over psychoactive substances following a war is recurrent in the history of the United States. Prohibition followed the First World War. Drug laws were tightened after World War II, and again following the Korean conflict. The severe penalties for the possession or sale of controlled substances in New York State in 1973 followed the addiction to heroin of many soldiers in the Vietnam War.

Legislated Law

As the old century waned and the new began, the Prohibition forces gained momentum and zeal. In 1906 the Anti-Saloon League was founded. A number of innovations in the treatment of alcohol dependence were tried. Among them was treatment through the use of heroin, which had been first synthesized in 1874. It was touted as a miraculous cure, as the addiction to alcohol was transferred to the narcotic. In Europe, at this time, Freud was advocating the use of cocaine to advance mental powers. The popularity of Freud's notion can be seen in the use of the drug by the absolutely analytic and totally controlled detective, Sherlock Holmes. Patent medicines abounded, many of which consisted in greater or lesser amounts of laudanum and alcohol. The Coca-Cola concoction introduced in this era enjoyed great popularity. Until 1903 its first name represented the cocaine that it contained. It now carries decocainized kola nut extract.[9]

The ease of obtaining laudanum potions or elixirs of dubious composition (many of which had an alcohol base) led to the inadvertent reliance of large numbers of the population on drugs. As the temperance forces were building steam for a national proscription against alcohol use, an estimated 215,000 persons were addicted to drugs in the United States.[7]

Although laws prohibiting the sale of alcoholic beverages had been passed in individual states, Prohibition made its first national appearance as a temporary measure enacted during World War I for the ostensible purpose of saving grain. A constitutional amendment prohibiting the "manufacture, sale, transportation, importation or exportation [of] intoxicating liquors" was passed by the Congress, December 17, 1917, while the war was raging. The necessary 32 states had ratified the amendment by January of 1919 and it went into effect a year thereafter, January 16, 1920.

The year's grace period was a slight recognition that a habit shared by much of mankind for several thousand years was to be denied. The Eighteenth Amendment is a marvelous testament to the idealism of the American people at this time. Having just fought and won the war "to make the world safe for democracy," a majority of those Americans taking the opportunity to vote on the issue believed they could solve many of the remaining social ills at home with the passage of a law proscribing "intoxicating liquors." Notwithstanding the status of the law as an amendment to the document that is the very foundation of our system of government, the Constitution, this was a heavy charge for three paragraphs of text. Unless this position truly represented the will of the majority of the residents of the country, including the disenfranchised poor and immigrant groups, the law could not work.

The Volstead Act, which provided flesh on the bare bones of the Eighteenth Amendment, was passed October 28, 1919. It provided the machinery for the enforcement of the law and the basis for the largest black market in the history of the country.

Judge-Made Law

Not surprisingly, and in keeping with the theme of this chapter, a movement parallel to Prohibition was occurring in relation to drugs. While Prohibition came about through the open, clearly visible process of constitutional amendment, the limitation on drug use grew by the equally legitimate, traditional common law process of court interpretation. Unlike the constitutional amendment route, when a rule is developed by judicial interpretation, its ultimate shape is not always clear from the beginning. This is not to say, however, that such a final determination is an ineluctable mystery until the court makes its final pronouncement. Construing the law and making judgments on its future interpretation, is one of the basic crafts of the lawyer.

An examination of the cases decided by the Supreme Court in this connection undermines the popular notion that the country's drug policy was the creation of a handful of isolated, power-hungry bureaucrats in the United States Treasury Department. This latter view* is set forth by Lindesmith in his book *The Addict and the Law:*

> The present program of handling the drug problem in the United States is, from the legal viewpoint, a remarkable one in that it was not established by legislative enactment or by court interpretations of such enactments. Public opinion and medical opinion had next to nothing to do with it. It is a program which, to all intents and purposes, was established by the decisions of administrative officials of the Treasury Department of the United States. After the crucial decisions had been made, public and judicial support was sought and in large measure obtained for what was already an accomplished fact.
>
> Another unusual feature of the federal narcotic laws is that, while they are in legal theory revenue measures, they contain penalty provisions that are among the harshest and most inflexible in our legal code.[10]

While the Treasury enforcement officials certainly had an effect, the fact is that the basis of narcotics policy in the early years of the century was the Harrison Narcotics Act, enacted December 17, 1914, and effective the following March.[11] The law appeared to be a relatively innocuous revenue measure on its face.† It required "every person who produces, imports, manufactures, compounds, deals in, dispenses, sells, distributes, or gives away opium or coca leaves or any compound, manufacture, salt, derivative, or preparation thereof" to

---

*To follow down the legacy of this concept, see Epstein EJ: *Agency of Fear.* Putnam, 1977.
†It will be remembered that seemingly innocuous revenue measures have caused great consternation in the past. The Stamp Acts imposed by the British Parliament on the colonists and the 1791 Whiskey Excise Tax are examples.

register with the federal government and pay a special tax of one dollar. The dollar tax was hardly onerous when compared with the complete prohibition of alcoholic beverages. Nonetheless, as was noted by Justice Marshall in *McCulloch vs Maryland* (4 Wheat. 316 [1819]), the power to tax is the power to destroy. Nor was there any doubt about the intention of Congress in creating the act.

The legislative history demonstrates clearly that Congress was seeking a means to control the traffic of narcotics in the country. The House of Representatives in its report on the bill stated flatly, "There is a real, and one might say, even desperate need of federal legislation to control our foreign and interstate traffic in habit-forming drugs, and to aid both directly and indirectly the States more effectually to enforce their police laws designed to restrict narcotics to legitimate channels."[12] The report went on to compare the 133% increase in the country's population between 1870 and 1909 with the 351% increase in imported opium in the same period[12] while the "almost shameless traffic in...drugs" was decried.[12(p 4)]

Exempted from the law were physicians, dentists, veterinary surgeons, and pharmacists acting in the course of their professional practice. This exemption and the advertising of the act as a revenue measure described the battlelines for the next ten years of litigation.

The first case requiring construction of the Harrison Act to come before the United States Supreme Court was that of Jin Fuey Moy, a Pittsburgh physician. Jin Fuey Moy had issued a prescription for one dram of morphine sulphate, not for medicinal purposes, but for the purpose of supplying one Willie Martin, a morphine addict. Treasury agents obtained an indictment under the Harrison Act against the doctor for "conspiracy to possess opium and the salts thereof."[13] The government in its case claimed that the Harrison Act was not simply a revenue measure but supported the exercise of police power in executing the international opium convention of 1912.

Justice Oliver Wendell Holmes in the opinion of the court held that the treaty did not compel the passage of such a statute. Furthermore, the criminal penalties set forth in the statute applied only to those who were required to register, that is, those who "import, produce, manufacture, deal in, dispense, sell or distribute." Since narcotic addicts did not have to register, it was not illegal, under the terms of the revenue act, to write a prescription for an addict. While the reasoning may appear a bit tortuous, the decision held the statute within its own terms. Justice Holmes, whose father was a physician, had maintained the sanctity of the medical judgment. The Harrison Act, by its terms, was not designed to second-guess a physician's prescribing practices.

In its next appearance before the Supreme Court, the Harrison Act was held to be a constitutionally enacted revenue measure and not an improper attempt to exercise the police power delegated by the Constitution to the states. The case portrays the marvelous ability of a law to grow to meet the social need, and as such represents the genius of the common law system. The matter concerned C.T. Doremus, a West Texas physician. The indictment alleged that Dr Doremus sold to one Ameris 500 one-sixth-grain tablets of heroin, "the sale not being in pursuance of a written order on a form issued on the blank furnished for that purpose by the Commissioner of Internal Revenue."[14] The second count recited the sale and stated that it was "not in the course of the regular professional practice of Doremus and not for the treatment of any disease from which Ameris was suffering, but, as was well known by Doremus, Ameris was addicted to the use of the drug as a habit, being a person popularly known as a 'dope fiend,' and that Doremus [sold] the heroin to Ameris for the purpose of satisfying his appetite for the drug as an habitual user thereof."[14]

The notion behind the Harrison Act, it will be recalled, was to provide a federal tool to be applied against the interstate and international traffic in narcotics. The states were presumed to be able to control internal distribution of narcotics through their own prescription laws. Indeed, many had. The territory of Nevada had outlawed opium dens and the sale of opium and its derivatives without a physician's prescription as early as 1877.[15] Yet, apparently the law in west Texas in 1918 was not able to control the sale of heroin by doctors and addicts. Neither did the strict terms of the Harrison Narcotics Act. The statute carried an exemption for "the dispensing or distribution of any of the aforesaid drugs to a patient by a physician, dentist, or veterinary surgeon registered under this Act in the course of his professional practice only."[16] By upholding the constitutionality of the law as a revenue measure in the Doremus case, the Supreme Court tacitly acquiesced in the creation of the crime of distributing narcotics to an addict.*

The social cause for the creation of the new crime is demonstrated in a case decided the same day as Doremus.[17] It shows how the court weighed the freedom of physicians to prescribe according to their own medical judgment versus the societal need for control. The facts were certified to the court as follows:

> Webb was a practicing physician and Goldbaum a retail druggist in Memphis. It was Webb's regular custom and practice to prescribe morphine for habitual users, upon their application to him therefor. He furnished these 'prescriptions', not after consideration of the ap-

*Four members of the court recognized this and cried "foul!" In their dissent they declared that such a construction of the statute is to allow "Congress to exert a power not delegated...the reserved police power of the states." (*United States vs Doremus,* pp 95, 217.)

> plicant's individual case, and in such quantities and with such direction as, in his judgment, would tend to cure the habit, or as might be necessary or helpful in an attempt to break the habit, but with such consideration and rather in such quantities as the applicant desired for the sake of continuing his accustomed use. Goldbaum was familiar with such practice and habitually filled such prescriptions.[17(97.217)]

The facts go on to establish, "It was the intent of Webb and Goldbaum that morphine should thus be furnished to the habitual users thereof by Goldbaum and without any physician's prescription *issued in the course of a good faith attempt to cure the morphine habit.*"[17(98.217)] Of course, the italicized requirement does not appear in the wording of the statute. The expansion of the statute in such a manner could and did bring into question the legitimacy of some 40 clinics across the country then providing treatment to addicts by maintaining them on opiates.[18] The facts, however, demonstrate that the Webb/Goldbaum connection was not overseeing a maintenance regimen. Rather, they were reaping significant monetary reward by selling opium to addicts. In 11 months Goldbaum sold narcotics in 6500 instances. His purchases from wholesalers were more than 30 times as much as the average retail druggist in Memphis doing a larger general business than the defendant. Webb, within the same period, had issued 4000 prescriptions for narcotics, charging fifty cents a piece for each. One Rabens came from another state and Webb issued him ten prescriptions at one time for one drachma each upon Rabens's representation that he was addicted. Goldbaum filled all of Rabens's prescriptions at one time even though they were made out in a separate and fictitious name.[17(98.218)]

The court held that Webb was not insulated by the physicians' exemption in the Harrison Act because prescriptions made out in this manner could not be deemed to be physicians' prescriptions. The court's reasoning is creative. It is not admirable legal scholarship, and it makes no effort beyond a recitation of the facts to justify its removal of the exemption.* Apparently the court felt that the practices of W.B. Webb and Jacob Goldbaum were so inimical to the social order that they deserved no protection even though the statute on its face provided it. Had not the prescribing practices been so heinous, the court would have had no foundation for its creative interpretation. By failing to regulate the outlandish practices of its members, the medical profession endangered the prescribing rights of all physicians.

The case in the following year once again involved Jin Fuey Moy.[19] He was the same Pittsburgh physician whom Justice Holmes had gotten off the hook in 1916. It would seem that the Treasury Depart-

---

*The decision was again 5 to 4 with the same dissents as in Doremus.

ment was acting out of pique at its initial setback and was persecuting the doctor. A look at the facts is instructive. Jin Fuey Moy was not issuing prescriptions on the required Internal Revenue Service blanks which he would have had to purchase and have the number he was issued recorded. His prescriptions were all to be filled at one Pittsburgh pharmacy, which filled them at the rate of "hundreds per month" throughout 1915, 1916, and 1917. He charged not by the visit nor even by the number of prescriptions but at one dollar per dram prescribed. He issued prescriptions "not in the ordinary course of professional practice...to persons not his patients and not previously known to him.... In some cases he made a superficial physical examination; in others, none at all."[19(193,100)]

The court held that issuing prescriptions in this manner was aiding and abetting a person to commit a criminal act, ie, dealing in narcotics without being registered. Although Jin Fuey Moy was registered, his patient-addicts were not. The case was decided without dissents. Many of the judicial giants of the century were on that court—Oliver Wendell Holmes, Louis D. Brandeis, William Howard Taft—but none of them registered any worry that they had participated in the creation of a new crime by judicial interpretation. Indeed, judge-made law enjoys a long and honorable history, and their act was by no means unique.

The next egregious act that demanded the removal of a physician's prescription from its otherwise exempt status under the Harrison Act was that of Morris Behrman, a New York City physician.[20] He wrote a prescription for Willie King, a person addicted to the habitual use of morphine, heroin, and cocaine. The impropriety of the act centered not so much on the act of maintenance itself, but on the amount supplied—150 grains of heroin, 360 grains of morphine, and 210 grains of cocaine.* Plainly, this was an indiscriminate distribution of narcotics and not a reasoned plan of medical treatment. The fact that the amount was ordered pursuant to a prescription did not fit within the exemption because the prescription was not within "the appropriate bounds of a physician's professional practice [but rather was] intended to cater to the appetite or satisfy the craving of one addicted to the use of the drug."[20(288,304)] The determination, however, was not unanimous. Justices Holmes, Brandeis, and McReynolds dissented on the grounds that the indictment made no declaration claiming that the prescribing was not done in good faith.[20(290,305)]

---

*According to Wood's United States Dispensatory, the ordinary dose of morphine at the time was one-fifth of a grain, of cocaine one-eighth to one-fourth of a grain, and of heroin one-sixteenth to one-eighth of a grain. Under these standards, more than 3000 doses were made available to King (*United States vs Behrman,* pp 288, 305).

The outlandish practices of Doremus, Webb, Jin Fuey Moy, and Behrman caused the court to invent a crime not specified by the Harrison Act. A reading of the cases shows that they were aimed at the most abusive of practices. However, Treasury Department officials, with the backing of the American Medical Association, interpreted the decisions to forbid any narcotic maintenance treatment. This overbroad interpretation, coupled with the characteristic caution of the medical profession in construing the law, led by 1925 to closing of every one of the public clinics offering a treatment regimen of narcotics maintenance.* Addicts who had been receiving their drugs legitimately were forced to turn to the black market to satisfy their needs.

The degree to which the control of narcotics had gotten out of hand was recognized in the case of *Linder vs United States.*[21] Dr Charles O. Linder of Spokane, Washington, was convicted of supplying one tablet of morphine and three tablets of cocaine to Ida Casey, who was addicted to the drugs.†[21(11)] In the Supreme Court Dr Linder's lawyer argued persuasively that the courts had "engrafted" an additional requirement on the exception from the act of a "physician, dentist, or veterinary surgeon registered under the act in the course of his professional practice only." The expansion was that "the drugs must have been dispensed or distributed in good faith as medicine, and not to satisfy the cravings of an addict.[21(6)] In other words what the court had been saying all along was not that prescribing narcotics was a crime per se, but that it was a crime when it was not done as part of a good faith course of medical treatment. The law was not to be read to make the poor and pathetic Ida Casey turn to the black market for her needs, but it was meant to stop the creation of a black market by the excessive and loose prescribing habits of Drs Behrman, Webb, and Jin Fuey Moy, and their other financially motivated colleagues. The argument was successful and Justice McReynolds, a dissenter throughout the previous string of cases, wrote a unanimous opinion stating that the Harrison Act "says nothing of addicts and does not undertake to prescribe methods for their medical treatment."[21(18)] He went on to explain that the Behrman case:

---

*National Commission on Marihuana and Drug Abuse, Second Report, 1973, p 308. The clinics were scattered across the country from Shreveport, Louisiana through Knoxville, Tennessee. They were not all in major urban centers. Clinics in New York existed at one time or another in Albany, Buffalo, Newburgh, Oneonta, Port Jervis, Rochester, Saratoga Springs, Binghamton, Corning, Hornell, Kingston, Middletown, Syracuse, Utica, and Watertown, as well as New York City. (Report of the New York State Joint Legislative Committee on Narcotic Study, Albany, 1959, p 18.)

†Again the pervasiveness of the use of narcotics at the time is demonstrated by the varied regions of the country from which the cases come. They are not all from urban ghettoes or brawling seaports.

> cannot be accepted as authority for holding that a physician, who acts *bona fide* and according to fair medical standards, may never give an addict moderate amounts of drugs for self-administration in order to relieve conditions incident to addiction.
>
> ...The unfortunate condition of the recipient [Ida Casey] created no reasonable probability that she would sell or otherwise dispose of the few tablets entrusted to her; and we cannot say that by so dispensing them the doctor necessarily transcended the limits of that professional conduct with which Congress never intended to interfere.[21(22)]

Looking back over these cases it can be seen that a crime was created by the judiciary, with a revenue act as its basis, in response to the abuses of the medical profession. The crime was then shaped to apply only to the abusers and not physicians practicing medicine in good faith and trying to provide poor Ida Casey a life of less pain. Unfortunately for the treatment field, the Treasury Department had succeeded, by the time the doctrine had been fully developed, in shutting down all the maintenance clinics.

The parallel to Prohibition again appears. Even though the sale of alcoholic beverages was prohibited generally, physicians were allowed to prescribe it for medical purposes.*

THE TREATMENT/PUNISHMENT DICHOTOMY

Drugs

Attempts to determine the wisest and most efficacious means of controlling alcohol and drug use have seesawed back and forth between punishment and treatment throughout the twentieth century. As for drugs, the response has been largely penal. Notwithstanding Prohibition, the response to alcohol has been less punitive. The reason for this disparity may not be entirely logical.

When the Harrison Act gained passage as a revenue measure, it was designed to supplement state laws controlling the use of drugs. A look at the development of the law in one state illuminates the dichotomy. In New York, laws regulating the prescribing of opium and morphine had existed as early as 1886.† Concurrent with the Harrison

---

*Indeed alcohol is still considered medicine in some quarters. In Connecticut, for example, when the "package" stores are closed, alcoholic beverages can still be obtained, by prescription, from a drug store!

†C. 390, Laws of 1886 at 610 made it a misdemeanor to "sell, give away, dispose of or offer for sale" opium or morphine without a scarlet label describing in white letters the name and address of the seller as well as the contents of the preparation. In 1887 pharmacists were prohibited from refilling prescriptions for opium or morphine more than once without a specific order from a physician to do so. (C. 636 Laws of 1887 at 848.)

Act, New York passed the Boylan Law.[22] It prohibited opiate medications except for the treatment of disease, injury, or deformity. Prescriptions were required for dispensing, and they had to be written on order blanks obtained from the State Health Commission. The names of all persons treated with opiates were required to be kept on record for five years. For the first time the concept of illegal possession of narcotics entered the law. The purchase and sale of syringes and needles were regulated. In addition to these proscriptions, treatment for narcotics addiction was authorized at hospitals. As with the Harrison Act, there was a great swirl of controversy as to whether the law meant that physicians were forbidden to administer and prescribe these drugs to their patients. It did not. Many physicians accepted the restrictive interpretation, however, and as a result great numbers of addicts crowded into hospitals that were not equipped for their treatment. The overflow sought to supply their addictions in the illicit markets of the street, and police blotters swelled with arrests of addicts.[23]

The First Whitney Act of 1917 loosened the restrictions on physicians and other medical personnel.[24] Furthermore, it permitted local boards of health to prescribe and dispense drugs free to addicts pursuant to regulations set by the State Board of Health. By 1918 the treatment approach to the drug problem was ascendent. Narcotic addiction was held to be a "disease" by a Joint Committee of the New York State Legislature.* The First Whitney Act plainly allowed drug maintenance in permitting physicians to prescribe for the purpose of "relieving stress"—an apparent euphemism for the symptoms of withdrawal. Under it and the Second Whitney Act, which created the Department of Narcotic Control to regulate them, maintenance clinics proliferated. Riverside Hospital in New York City enjoyed its first use as a detoxification center during the period. But in their natural caution in construing law, physicians felt that the euphemism of "stress" was not a clear enough authorization to protect them. Furthermore, they objected to regulations promulgated by the Department of Narcotic Control. The result was a stormy legislative session in 1921 which passed several major bills relating to drug treatment, including a provision to repeal the Whitney Act. The governor vetoed all but the last. New York was without significant drug control legislation through 1927.[15(713)]

---

*And a prevalent disease at that; the committee judged that up to 5% of the New York City population was addicted. Further, narcotics addiction was "more prevalent and widespread in the smaller cities and rural communities than has been believed to be possible." Final Report of Joint Legislative Committee to Investigate the Laws in Relation to the Distribution and Sale of Narcotic Drugs, New York State Senate Report No. 35, 1918, p 3.

The discomforture of physicians with the law, notwithstanding the continuing authority of the Harrison Act, led to the closing of narcotic maintenance clinics not only in New York but across the nation. The response of the states "shifted to a strictly penal approach."[7(50)] The result was to bring on 50 years of attempting to deal with the drug problem by punishment rather than any serious regimen of treatment.

Notwithstanding the stringency of these measures, there was no appreciable drop in addiction from 1925, when the production and importation of heroin was prohibited, until World War II.[7(50)] During that period most states adopted laws prohibiting the possession and use of narcotics. The shift to the penal approach is remarkable when seen against the backdrop of national Prohibition. Even though the Constitution now prohibited the "manufacture, sale, or transportation of intoxicating liquors," only five states* saw fit to prohibit the private possession of alcohol for personal use.[25] Even as Congress was passing the Twenty-first Amendment repealing Prohibition, the states were adopting the Uniform Narcotic Drug Act. That law, which was adopted in some form by all states except New Hampshire, set forth prescribing and recording procedures and regulated the manufacture and sale of narcotics. Places where narcotics were dispensed illegally or where narcotics addicts congregated were declared public nuisances.[26]

Apparently drugs were perceived in the public eye as far more insidious than alcohol. It seemed to be felt that the use of narcotics led to a life of doom and disaster. Usage was causally linked to crime. On the other hand, it was thought that alcohol could be used in moderation while the social fabric was maintained. Alcohol use was deemed controllable by regulated sales and restrictions against consumption by minors.

The weight of scientific evidence is not wholly in support of these public perceptions. In the mid-1920s a well-respected psychiatrist was arguing that neither a heroin epidemic nor the total elimination of opiate addiction would make an appreciable impact on the overall amount of crime.[27] The only effect of heroin was, he felt, to make the addict less of a murderer and more of a thief. Later studies back up this view. Isadore Chein found in 1964 that the increase in property crimes by juvenile addicts of the early 1950s was offset by the decrease in violent crime.[27] It is the conclusion of a recent study that "given the present state of research, there is no reason to believe that addiction is the crucial variable which accounts for increases in criminality."[27(244)] That study goes on further to state that "the weight of the evidence suggests that the probability of violent behavior is not substantially in-

*They were Georgia, Idaho, Indiana, Kansas, and Tennessee.

creased by heroin abuse."[27(245)] If it is increased at all, it is not during the period of euphoric nod produced by the injection, but more likely at the onset of withdrawal, when the illegality of the drug makes it necessary for the addict to commit crimes to obtain the money necessary for a purchase on the illicit market.

The penal response of the 1920s and 1930s was not the only instance of a draconian approach to the social problem of drug abuse. Rather, it represents the propensity of legislators confronted with a crime problem to conclude that "the best hope of *control* lies in 'getting tough' with criminals by increasing penalties."[28] It is a phenomenon that has been observed "to be particularly apparent in the area of regulation of dangerous drugs."[28]

The punitive approach all but eclipsed the treatment response to drug addiction from the mid-twenties through the 1950s. Perhaps as an anachronism, New York continued to carry on its books a 1927 law* that allowed a judge to commit to a private licensed institution for the insane a person:

> over the age of eighteen years [who] is incapable or unfit to properly conduct himself or herself, or his or her affairs, or is dangerous to himself or herself or others by reason of periodical, frequent or constant drunkenness, induced either by the use of alcoholic or other liquors, or of opium, morphine, or other narcotic or intoxicating or stupefying substance.[29]

The lack of distinction in the commitment statute between drunkenness caused by alcohol and drunkenness caused by narcotics or other "stupefying substances" is noteworthy. The purpose of the commitment was therapeutic. Commitment was based on a showing that the person was in "actual need of special care and treatment, and that his condition is such that his detention, care and treatment in such *institution* would be likely to effect a cure."†

During the period the laws of 34 states provided for treatment of drug-dependent persons in mental hospitals.[25(310)] In New York and California some persons actually received treatment in these institutions, but the effort was not large. The bulk of drug-dependent persons receiving treatment got it at the two federal "narcotic farms." The first of these was opened at Lexington, Kentucky, in 1935. In 1938 the second opened in Fort Worth, Texas. There were more than 10,000 admissions to treatment during a 30-year period, with many of them

---

* The law is unusual in its sex-specific pronouns. Even today with significant advances in sexual equality the standard form for legislative drafting is the theoretically all-inclusive male pronoun.

†Notice the reversion to the male pronoun.

being readmissions of former patients.[25(310)] While voluntary admissions were allowed initially in the operation of these federal prison-hospitals, it was found necessary later on to accept only patients convicted of a drug offense. The penal emphasis prevailed.

It was not until 1952 when Riverside Hospital reopened to receive addicted patients that there was any significant state effort to supplement the federal treatment facilities.[30] By 1959 it was still the only hospital of its kind in the world.

While many states had statutes on their books calling for treatment of drug addiction, few of them made funding available in sufficient amounts to support the service. Nor did the public hospital system make the necessary care available. The case of New York City is reflective of the general attitude of the times.

In 1958, as it had been for decades, the policy of the New York City Department of Hospitals was not to admit drug addicts. Drug addiction was the only disease category specifically excluded from the city hospitals according to Dr Henry W. Kolbe. As the reason for denying treatment Dr Kolbe cited the "lack of knowledge about addiction, the effect of drugs and the desirability of withdrawing persons from the use of drugs."[23(39,41)] An appropriation of $160,000 for a pilot research project on the treatment of adult narcotic addicts went unused for two years because none of the medical school hospitals in the City of New York was willing to undertake such a project. The result was that addicts seeking detoxification had to voluntarily commit themselves to prison to obtain treatment.[23(39,41)] This occurred at a time when hospitals in three of the five boroughs provided special wards for alcoholics.

The real activity in the law relating to drugs was on the penal front. The post-World War II–Korean War period brought a drastic increase in the penalties for narcotic offenses on a nationwide basis.[31] For the first time, the mere possession of a narcotic without proof of any intent to use it illegally was made a crime. In New York the penalty for possession of narcotics became the same as for sale—ten years.[32]

After the Korean War, penalties were again tightened. The federal law was amended to require mandatory minimum sentences without probation or parole for all but first offenders.[33] State legislatures followed suit.[34] Within a decade the Federal Drug Abuse and Control Amendments of 1965 increased penalties once again.[35] Included in these amendments was the comprehensive Depressant and Stimulant Drug Control Act regulating the use and sale of amphetamines and barbiturates.[35] In New York State, hallucinogenic drugs were also brought under control.[36]

Penalogical fervor was so strong at this time that the United States Supreme Court was compelled to declare that it was not a crime simply to be an addict. The City of Los Angeles had adopted an ordinance that made it a misdemeanor to "use, or be under the influence of or be addicted to the use of narcotics." The court* reminded the nation that narcotic addiction was a disease, and that "in the light of contemporary human knowledge, a law which made a criminal offense of such a disease would doubtless be universally thought to be an infliction of cruel and unusual punishment in violation of the Eighth and Fourteenth Amendments."[37]

This reminder seemed to set the stage for a ride back on the pendulum to the treatment response to drug abuse. New York had included in its penal law a provision allowing narcotic addicts to undergo treatment as an alternative to incarceration or as a condition of parole or probation.[38] California and New York both passed extensive provisions for the diversion of addicts from the criminal justice system.[39] Synanon and other therapeutic communities for the treatment of drug addiction were formed. By 1966 New York had created the Narcotic Addiction Control Commission to provide treatment services to persons suffering from drug dependence and abuse. By 1971 the agency was operating 16 secure treatment facilities in the state with a patient population approaching 4000. Additional treatment facilities were provided by state funding of local community-based treatment programs, including residential, day service, and outpatient drug-free programs, as well as detoxification and methadone maintenance programs.[40] California and Massachusetts had similar approaches calling for the voluntary and involuntary commitment of drug-dependent persons to treatment.[41] By the mid-1970s, 18 states had laws adopting the New York–California approach.[42]

A major overhaul of the federal legislation on control of dangerous drugs was achieved in 1970 with the passage of the Controlled Substances Act. The act categorized the dangerousness of more than 170 drugs into five different schedules and set specific requirements for the dispensing of each.[43] The reorganization of the federal system of control compelled the states to abandon the Uniform Narcotic Drug Act under which they had operated since the 1930s and supplant it with some variant of the Uniform Controlled Substances Act.[44]

The structure of the law relating to the control and use of drugs has not been shaped solely by the dispassionate knowledge of science. The winds of political opportunism have joined in the sculpting as well. In the 1930s marijuana was hyped as a "killer weed" by certain federal enforcement officials in order to advance their own personal positions.

---

*Actually the case should not have been decided by the court under the doctrine of mootness. Robinson was dead by the time the court made its decision—apparently of an overdose of narcotics.

Edward Epstein perceives an attempt to create a national police force during the Nixon years in the guise of a drug control agency.* One of the starkest examples occurred in New York in 1973. There, the pendulum was well on its way through the treatment part of its swing when it was suddenly arrested and reversed. The apparent cause was Governor Nelson Rockefeller's desire for the presidency. Feeling shackled by a liberal record in his pursuit of the 1974 Republican nomination, Governor Rockefeller sought ways to present a more conservative image. To this end he suddenly declared that the treatment response to drug abuse, which he had espoused and fostered, did not work. A new approach was needed; it was time to crack down on drug abusers with increased penalties, including mandatory life sentences with no chance for probation or parole.

The Rockefeller laws passed.[45] It appears their result has been to increase the costs to the judicial and correction systems without producing an appreciable decrease in the amount of drug use or drug-related crime.[46] Even now, as the effects of the mandatory sentencing provisions are resulting in the closing of the last state-operated treatment facility, there is a movement in the New York State Legislature to remove from its statute books the harshest drug penalties in the nation. The pendulum is on its way again!

Alcohol Abuse: Punishment or Treatment?

By the early 1930s the country had had enough of Prohibition. The Democratic Party plank of 1932 called for its repeal. In February 1933 Congress passed the Twenty-first Amendment to repeal the Eighteenth Amendment. Ratification by the necessary two-thirds of the states was swiftly achieved and, by December 1933, the nationwide prohibition against sale of alcoholic beverages was no more.

To parallel the original scheme of the Harrison Act, control over distribution of alcohol was now left to the states. The result was a patchwork of regulatory programs. Some states imposed or continued outright prohibition on their own. Pennsylvania and other states allowed over-the-counter sales only at state-run stores. Most states chose some form of licensure overseen by a state control commission. Various idiosyncratic local rules were generated by the overseers. In Washington, DC, until recently, one was not allowed to drink in a bar or restaurant while standing up. Until 1978 liquor could not be purchased by the drink at North Carolina restaurants, although diners could bring along their own bottle if they wished. Most states had some

*To follow down the legacy of this concept, see Epstein EJ: *Agency of Fear.* Putnam, 1977.

sort of Sunday closing rules, ranging from absolute prohibition of Sunday sales through the Philadelphia rule where only certain restaurants and hotels could sell on Sunday, to New York's rather lenient prohibition on beer sales between 4 AM and noon on Sundays. Of the few states which maintained a statewide prohibition, all had abandoned it for local option by 1966.

While the imbibing of alcoholic beverages continued to be deemed a sin in certain narrow quarters, it achieved general acceptance and even some status over most of the country. Public advertising was allowed. Practitioners of the art of convincing the public mind have sought to link certain brands with social success and financial achievement. Use of alcohol and even lower levels of abuse are generally tolerated.* Nonetheless, the societal response to alcohol dependence is to some degree parallel to the response to drug use. The duality between punishment for the excessive use and treatment for the disease of alcoholism is again manifest.

Public drunkenness has a long history of criminal sanction. It was explicitly proscribed by English statute in 1606.[47] In 1968 it was an offense in every state of the union. Yet the efficacy of the penal proscription has long been questioned.[48] By 1965 one-third of all arrests† made in the United States were for the crime of public intoxication.[49] In the 1960s the procedure of jailing for public drunkenness came under attacks in the courts. The arguments sprang from the Supreme Court's holding in Robinson that the status of being a narcotic addict was not an offense.[37]

On that basis, the Federal Court of Appeals for the Fourth Circuit held that it was unconstitutional to punish public drunkenness. It was cruel and unusual punishment to send a person to jail simply for being in a state of intoxication.[50] The District of Columbia Circuit Court, which was playing an activist role in the development of mental health law throughout the 1960s and after, again declared it unconstitutional under the Eighth Amendment to punish a homeless alcoholic who had been arrested 70 times for public intoxication.[51] Since the defendant was a chronic alcoholic who by the nature of his affliction could not resist drinking, his act of appearing in public in a state of intoxication was an involuntary act. Because of the involuntary nature of the act, it was without *mens rea,* the intent to commit a crime, which is necessary for conviction.

The issue reached the United States Supreme Court in the case of *Powell vs Texas.*[52] Leroy Powell was a chronic alcoholic who eked out a

---

*Mark Twain is reported to have remarked that "too much of anything is too much, but too much whiskey is just right."

†This amounted to more than two million arrests.

living by shining shoes for $12 a week in an Austin bar. He had been convicted in the Travis County Court of violating the article of the Texas penal code that stated:

> Whoever shall get drunk or be found in a state of intoxication in any public place, or at any private house except his own, shall be fined not exceeding one hundred dollars.[53]

The trial court had rendered a guilty verdict couched in the terms of the Robinson decision, thereby appearing to offer the Supreme Court the opportunity to throw out the Texas law as unconstitutional. The Supreme Court sidestepped the Robinson argument, although not without dissent. There was no majority opinion in the case. Justice Marshall wrote the plurality opinion in which three justices joined, one of those three writing a separate, concurring opinion of his own. Three other justices dissented.

The court held that there was a distinction between the status of being a narcotics addict which the California law had punished in Robinson and the act of appearing drunk in a public place.[52(532.2154)] Two liberal concerns seemed to motivate Justice Marshall in arriving at his distinction. One was the unavailability in most regions of the country of sobering-up stations as an alternative to jail.[52(528.2152)] The other was a fear advanced by the American Civil Liberties Union as *amicus curiae.* Justice Marshall shared the fear that relatively short sentences of a night or so, which were simply designed to allow the defendant time to sober up, might be replaced by an open-ended civil commitment for the condition of alcoholism. The opinion stated:

> One virtue of the criminal process is, at least, that the duration of penal incarceration typically has some outside statutory limit; this is universally true in the case of the petty offenses, such as public drunkenness, where jail terms are quite short on the whole. 'Therapeutic civil commitment' lacks this feature; one is typically committed until one is 'cured'. Thus, to do otherwise than affirm [the conviction] might subject indigent alcoholics to the risk that they may be locked up for an indefinite period of time under the same conditions as before, with no more hope than before of receiving effective treatment and no prospect of periodic 'freedom'.[52(529.2152)]

The court was discomforted that there seemed to be no clear consensus in the field on a definition of alcoholism, nor was there general agreement on the methods by which the condition could be effectively treated. Given this situation, the court declined to declare a constitutional rule for the treatment of alcoholism across the nation. Rather, the matter was one that could be more suitably handled by the individual states:

> Formulating a constitutional rule would reduce, if not eliminate, that fruitful experimentation, and freeze the developing productive dialogue between law and psychiatry into a rigid constitutional mold. It is simply not yet the time to write into the Constitution formulas cast in terms whose meaning, let alone relevance, is not yet clear either to doctors or to lawyers.[52(536,2156)]

Once again the job was left to the states. Their response has generally been the decriminalization of public intoxication. The year after the Powell decision, the Supreme Court of Minnesota held that alcoholism was a complete defense to a charge of public intoxication.[54] California devised the notion of "civil protective custody" for drunkenness.[55] The person detained thereunder is not to be charged with a penal or juvenile offense. Massachusetts passed an "Act Establishing a Comprehensive Program for the Treatment and Rehabilitation of Intoxicated Persons and Alcoholics and Abolishing the Crime of Public Intoxication" in 1971.[56] By 1975, 17 states had decriminalized their public intoxication statutes along the lines suggested by the Uniform Alcoholism and Intoxication Treatment Act.[57] A major reorganization of the Mental Hygiene Department in 1977 saw the removal of the involuntary commitment sections for intoxication from the Mental Hygiene Law of New York.[58] The dominant trend across the country appears to be the three-pronged approach adopted by the Uniform Act. The prongs include decriminalization, short-term detoxification, and some sort of civil commitment for the severely incapacitated.[59]

Alcohol and Drug Use Compared

As early as 1927 alcoholics and drug addicts were covered by the same commitment provisions of the law in New York State as well as elsewhere.[60] Involuntary commitments to licensed local asylums were provided for "inebriates," meaning persons who were incapable of handling themselves due to "periodic, frequent or constant drunkenness induced by the use of alcoholic or other liquors, or of opium, morphine, or other narcotic or intoxicating or stupefying substance."[60(§201)] No distinction was made between the stupor induced by drugs or alcohol.

With the relative legitimation of alcohol after the repeal of Prohibition, differentiation in the responses to the substances appeared. While the abuse of either substance generated a penal response by society, the difference in the level of that sanction is striking. Contrast the difference between the short jail stay described by Justice Marshall in the Powell case[52] to the prison terms standardly meted out for the basic drug offenses of possession and sale. The 1973 Drug Laws in New York* are an extreme example with mandatory life sentences, without probation or parole.[45] Even less extreme states commonly impose prison sentences of a year and more for the possession of small amounts of narcotic and other drugs for personal use.

The disparity is peculiar in that it seems to be born out of the contagion of fear about drugs rather than out of rational weighing of the dangerousness of each substance. The National Commission on Marihuana and Drug Abuse ranked substances for their propensity to produce violent behavior in the following order: alcohol, stimulants, hallucinogens, and finally opiates.[25(156)] It has been observed that the penalties under the laws of New York State and most states are "almost precisely inversely proportional to the degree of social danger they represent."[28(649)]

The California device of "civil protective custody"[55] applies only to persons under the influence of alcohol solely. Persons under the influence of a drug or a drug and alcohol may not qualify.

Social Security and Human Rights Law

The disparity in society's attitudes toward alcoholism and drug dependence is reflected in the developments of the case law on Social Security disability claims. Initially, alcoholism was viewed as "a voluntary condition, correctable by an act of will, and therefore, not within the purview of the disability insurance program."[61] However, recent cases have advanced the notion that not all alcoholics have made a "voluntary resignation to alcoholism as an escape of choice from a life of daily labor," but that in some instances alcoholism is a "helpless self-entrapment in an unconquerable addiction."[62] Thus, advanced levels of alcohol dependence have been determined to be qualifying conditions for Social Security disability claims. However, the mere presence of drug dependence without additional medical or psychological dysfunction is not deemed to be a qualifying disability.[63]

In New York State the Division of Human Rights has determined that alcoholism is a disease. As such it is illegal to discriminate against an alcoholic because of the disability.[64] Until recently, the Division was not willing to extend its protections to sufferers from drug dependence. In the case of *Perez vs State of New York,* Domingo Perez was denied a civil service position of elevator operator and building guard

*In New York drug felonies carry a higher designation and a longer prison sentence than arson or kidnapping second, or burglary, manslaughter, rape, and robbery at the highest level.

solely because he was receiving methadone maintenance treatment. He qualified on all other grounds. The New York State Human Rights Division dismissed Perez's claim that he was illegally discriminated against because of his medical disability by declaring that "drug dependence is a social problem, not a medical problem."[65]

The New York State Human Rights Appeal Board affirmed the dismissal of the complaint. However, the lawyers from the Legal Action Center, a public interest law firm representing Perez, were undaunted. Armed with a solid legal argument and supplemented by an *amicus curiae* brief from the New York State Division of Substance Abuse Services, the lawyers proceeded to the Appellate Division. The court annulled the decision and remanded the case to the Human Rights Division for further examination of the question.[66]

In an opinion from the general counsel, the Human Rights Division reviewed the legal developments in the area and reversed its position.[67] Henceforth, drug addiction is to be considered "a medical problem and therefore a disability pursuant to §292.21 of the Human Rights Law.[67(2)] Of course, if the drug dependence makes it impossible for a person to perform a specific job, it is not unlawful discrimination to deny employment on that ground. However, participation in a methadone maintenance treatment program does not in and of itself disqualify an addict from employment across the board.

Efforts to bring the law of discrimination into congruence for both alcoholics and drug addicts have not all been so successful. New York City is one of the most liberal jurisdictions in the country as well as being preeminent in its numbers of drug addicts and alcoholics. Nonetheless, the City Council has repeatedly refused to enact a local ordinance that would make it unlawful to discriminate against drug-dependent persons as it is against alcoholics. The decision of the State Human Rights Division has now effected the equal treatment of the conditions that the City Council was unwilling to impose.

The weight of medical evidence shows that the abuse of alcohol can be more damaging to the body than heroin and other addictions. The evidence from the criminal justice system shows aggressive behavior and violent crime to be more often associated with alcohol abuse than drug dependence.[25(165)] One can only speculate that the causes for the disparate views may relate to the fear of the unknown. Far more of the citizenry have had experience with alcohol and have used it on a regular basis than have experimented with illicit drugs. This, combined with the federal enforcement effort, which appears to have converted a general lack of concern with and ignorance about drugs into widespread alarm and misinformation, may have served to shape a social policy based more on fear than fact. The Perez decision represents a step toward a more rational approach to the problem.

IS WHAT IS PAST ALSO PROLOGUE?

In the ebb and flow that is the law of alcohol and drug abuse there is discernible movement away from the incongruities and toward the parallels. If New York City may be taken as a microcosm one last time, the development in recent years highlights the parallels. In preparation for the decriminalization of the public intoxication offense, a network of treatment and sobering-up facilities has been developed across the state. A similar network of more than 400 substance-abuse facilities has been created by the state. Some of the dragon's teeth have been pulled from the 1973 drug laws by the easing in 1979 of certain of the mandatory sentencing provisions.[68] The drive now seems to be toward treatment as an alternative to incarceration. Governor Hugh Carey has highlighted this movement in his State of the Health Message to the Legislature: "As an alternative to the automatic prison sentences which often in the past faced drug abusers, the Division of Substance Abuse had demonstrated that rehabilitation programs can be successful and that those under treatment can return to the labor force as productive members of the community."[69] As the involuntary commitment provisions of the mental hygiene law relating to alcoholics were dropped in the 1977 creation of the Office of Alcoholism and Substance Abuse Services,[58] so the Recodification of the Substance Abuse Treatment Law, pending before the legislature at the time of this writing, excises the involuntary commitment of substance-dependent persons.[70] The very creation of the agency combining both the alcoholism and substance-abuse efforts manifests a growing trend across the country. Who knows, maybe one day we shall return to the early notions that the problems of alcohol and drug abuse and dependence are parallel, if not identical, and should be dealt with in similar ways. After all, we have been there before.

REFERENCES

1. *Egyptian Temperance Tract,* circa 1000 BC, cited in "The public inebriate and the police in California," *Golden Gate Law Review* 5:259, 1975.
2. *The Odyssey,* Book IX, WHD Rouse translation, Mentor Books, 1949, pp 101–102.
3. Proverbs 27:9.
4. Schultes RE: Hallucinogens of plant origin. *Science* 163:250, 1969.
5. Blum, et al: *The National Experience.* 1963, pp 149–150.
6. Mitchell B: *Alexander Hamilton: A Concise Biography.* Oxford University Press, 1976, p 273.
7. Hardt CR, Brooks R: Social policy on dangerous drugs. *St. John's Law Review* 1973, p 48.

8. Berridge V: Working-class opium eating in the nineteenth century. *Br J Addict* 73:363, 1978.

9. Peterson J: The history of cocaine, in Peterson J, Stillman A (eds): *Cocaine: 1977.* National Institute of Drug Abuse, Monograph No. 13, 1977, p 17.

10. Lindesmith AR: *The Addict and the Law.* New York, Vintage Books, 1965, p 2.

11. Public Law 63-223; 38 Stat 785; 6 US Comp Stat 1916 § 6287g.

12. 63d Congress, Senate Report 258, *Registration of Persons Dealing with Opium,* February 18, 1914, p 3.

13. *United States vs Jin Fuey Moy,* 241 U.S. 394, 36 S.Ct. 658 (1916).

14. *United States vs Doremus,* 249 U.S. 86, 39 S.Ct. 215 (1919).

15. Quinn and McLaughlin, The evolution and present status of New York drug control legislation, *Buffalo Law Review* 22:705, 709, 1973.

16. Public Law 63-223 § 2(a).

17. *Webb et al vs United States,* 249 U.S. 96, 39 S.Ct. 217 (1919).

18. Mecher EM, eds of Consumers Union: *Licit and Illicit Drugs, 1972, p 115.*

19. *Jin Fuey Moy vs United States,* 254 U.S. 189, 41 S.Ct. 98 (1920).

20. *United States vs Behrman,* 258 U.S. 280, 42 S.Ct. 303 (1921).

21. *Linder vs United States,* 268 U.S. 5, 45 S.Ct. 446 (1925).

22. C. 363 Laws of 1914.

23. Report of the New York State Joint Legislature Committee on Narcotics, Albany, 1959, p 16.

24. C. 431 Laws of 1917.

25. National Commission on Marihuana and Drug Abuse, Second Report, 1973, p 244.

26. C. 684 of the Laws of 1933 is the New York rendition.

27. Lawrence Kolb, quoted in Crime and addiction—An empirical analysis of the literature, 1920–1973. *J Contemp Drug Problems* 3:221, 241, 1974.

28. Glanville RE: Drug abuse, law abuse and the eighth amendment. *Cornell Law Review* 60:638, 660, 1975.

29. C. 426, Laws of 1927 at section 201.

30. C. 8, Laws of New York 1952.

31. Act of November 2, 1951, C. 666, of the Laws of the United States, 65 Stat. 767; C. 529-530, Laws of New York 1951 amending the Penal Law.

32. C. 346, Laws of New York 1950.

33. Narcotic Drug Control Act of 1956, C. 629, 70 Stat. 567.

34. C. 526, Laws of New York 1956, p 1248.

35. P.L. 89-74, 79 Stat. 226.

36. C. 332, Laws of New York 1965, p 1973.

37. *Robinson vs California,* 370 U.S. 660, 763 82 S.Ct. 1417 (1962).

38. C. 526, Laws of 1956.

39. California Stat. 1961 850; C. 204, Laws of New York 1962.

40. New York Mental Hygiene Law. Articles 81 and 83, McKinney's, 1976.

41. California Welfare and Institutions code § 3000 et seq., Massachusetts Rehabilitation and Treatment Act for Addicts and Drug Dependent Persons, c. 889 Acts 1969; c. 123 § 38 et seq. Mass. General Laws eff. Jan. 1, 1971.

42. (Alabama) Ala. Code Title 22 §§ 249, 250; (Alaska) Alaska Stat. § 244-29-150; (Arizona) Orig. Rev. Stat. § 36-2001; (Arkansas) Ark. Stat. Ann. § 59-901 et seq; (Colorado) Colo. Rev. Stat. § 12-22-301 et seq; (Delaware) Del. Code Title 16 § 4801 et seq; (Florida) Fla. Stat. Ann. § 397 et seq; (Georgia) Ga. Code Ann. §§ 88-401 et seq; (Iowa) Iowa Code Ann. § 204.409 et seq; (New

Hampshire) N.H. Rev. Stat. Ann. § 172.13 et seq; (New Jersey) N.J. Stat. Ann. § 26-26-21 et seq; (Oklahoma) Okla. Stat. Ann. Title 43A § 651; (Texas) Tex. Code Crim. Proc. Ann. Art. 5561c-1; (Virginia) Va. Code §§ 37.1-220c.

43. 21 USC § 801 et seq.

44. 1970 Handbook of the National Conference on Commissioners of Uniform State Laws 225.

45. C. 276, 277, 278, 1051 Laws of 1973; Penal Law § 220 et seq.

46. Association of the Bar of the City of New York. Final Report on the Joint Committee on New York Drug Law Evaluation, 1977.

47. 4 Jac. 1 Ch 5.

48. Pittman A, Gordon B: *The Revolving Door: A Study of the Chronic Police Case Inebriate.* Glencoe, Ill, Free Press, 1958.

49. *President's Commission on Law Enforcement and the Administration of Justice Task Force Report: Drunkenness.* 1967, p 2.

50. *Driver vs Hinnant,* 356 F.² 761 (4th Circ. 1966).

51. *Easter vs District of Columbia,* 361 F.² 50 (D.C. Cir. 1966).

52. *Powell vs Texas* 392 U.S. 514, 88 S.Ct. 2145 (1968).

53. Texas Penal Code, Art. 477 (1952).

54. *State vs Fearou,* 283 Minn. 90, 166 N.W. 720 (1969).

55. California Penal Code § 647 ff. (1969).

56. Chapter 111B of the Massachusetts General Laws (Alcoholism Treatment and Rehabilitation Law 1971).

57. Alaska Stat. § 47.37.010 (1973); Fla. Stat. Ann. § 396.022 (Supp. 1973); Kan. Gen. Stat. Ann. § 65-4002 (Supp. 1973); Me. Rev. Stat. Ann. title 22, § 1361 (Supp. 1974); Md. Ann. Code art. 2C, § 102 (Supp. 1973); Mass. Gen. Law Ann. C. 111B, § 8 (Supp. 1974); Minn. Stat. Ann. § 340.961 (1972); Mont. Rev. Codes Ann. § 69-6211 (Supp. 1974); Nev. Rev. Stat. § 458.260 (1974); N.D. Cent. Code § 5-01-05.2 (Supp. 1974); Ore. Rev. Stat. § 426.460 (1974); R.I. Gen. Laws § 40.1-4-2 (Supp. 1973); S.D. Comp. Laws § 34-20A-1 (Supp. 1974); Wash. Rev. Code Ann. § 70.96A.010 (Supp. 1973); Hawaii Rev. Stat. R.L. 1945, § 11214 (repealed 1968); Ohio Laws 127 v. 1039, § 107 (repealed 1973); Laws 1973 C. 303, § 20-2, N.M. Laws 1963 (repealed 1973).

58. C. 978 Laws of 1977.

59. 5 Golden Gate Law Review at 268.

60. C. 426 of the Laws of 1927.

61. *Griffis vs Weinberger,* 509 F² 837 (Ninth Cir. 1975).

62. *Bardichek vs Secretary of Health, Education and Welfare,* 374 F. Supp. 940 (E.D.N.Y. 1974).

63. Social Security regulations at 20 C.R.F. 416.981.

64. *New York Law Journal,* August 22, 1978, p 22, col 3.

65. Div. Human Rights Complaint IVEDNR 256478 March 8, 1978.

66. 70AD2ᵈ558, 416 NY Supp. 2ᵈ813 (1st Dept. 1979).

67. Opinion of Ann Thacher Anderson, General Counsel, State of New York Division of Human Rights dated November 21, 1979.

68. C. 410 of the Laws of 1979.

69. NY Governor's State of the Health Message, February 27, 1980, p 24.

70. Assembly Bill 6955, 1980 session.

10 Political Aspects of Alcoholism and Drug Abuse

Kim A. Keeley

An orderly approach to naturally occurring events is the hallmark of professional inquiry. This tactic can be useful in sorting out the public and political events that impact the world of alcohol and drug abuse treatment.

This chapter deals first with stereotyping, a major phenomenon of the professional-political dialogue. The sense that the politician and the professional come from different worlds and speak different languages is so real and pervasive for both groups that this aspect of resistance to communication is considered at the beginning. The intent is not to make the resistance disappear so much as it is to introduce a basis for acknowledging and living with its reality. It should be permissible for politicians and professionals to mention their frustrations in dealing with each other, and the chapter provides a way of viewing this frustration along with any subsequent satisfactions. Included here are some standard reasons either group has for ignoring the other or for reaching out to collaborate.

After these familiar motivations are clarified, interest is focused on the types of data from the world of politics that must be handled by the alcohol or drug abuse professional. In a political setting where all events appear to be contingent upon one another, there is every temptation to ascribe causality to situations that, in reality, lie outside the immediate control of most politicians and professionals. To overcome such appearances, these two need a frame of reference that permits them to discriminate relevant from not-so-relevant information about each other.

For the purposes of the alcohol and drug abuse professional, thinking of politics as a behavioral system is a useful paradigm, and the languages of systems analysis, psychology, and psychiatry provide as comfortable modes as any to describe its functions. Politics as a behavioral system gets its input from a variety of sources, which can be characterized by the content of the data input as well as the organizational complexity of the data source. These variables are of extreme interest to the political person, although political people would hardly call them variables. The health professional can use a list of these variables to become oriented to the facts of life as the politician sees them.

The final portion of this chapter identifies some of the tasks of the politically oriented professional and the professionally oriented politician, especially those who may be leaders in the field of alcoholism and drug abuse. Just as research scientists ask questions about the basic science of substance abuse and clinical scientists can help determine whether and when research advances become available to patients, professionals with political leanings also have opportunities to shape their world. This professional acts on a stage which may be viewed either large or small, the dimensions often depending on whichever way one chooses to see them.

Finally, it is worthwhile identifying some of the great issues that energize the political dimensions of alcoholism and drug abuse. There is often a critical role to play in the calibration of professional and political vectors and in the balancing of collective versus individualistic health themes. These roles bring with them an increased likelihood that the frustration that professionals often feel about politics and the frustration that society often feels about chemical dependence can be transformed via specific achievements into a certain measure of satisfaction about having spent some time or money working in this field.

MUTUAL STEREOTYPING BY THE PROFESSIONAL AND THE POLITICIAN

Differentiative Bonding[1] Themes

It is clear that health professionals, whether in or out of the drug and alcohol fields, and politicians can rightly claim that they are ignored by each other. Organized medicine believes it should be consulted more actively when politicians develop national health insurance plans. Conversely, government officials regularly claim that health professionals have little incentive to curtail spiraling costs.

While these economically oriented examples are readily grasped by contemporary society, it is just as true that political and professional groups ignore each other when a variety of salient issues are concerned. For example, professionals pay only temporary attention to an elected official's impression of a scientific hypothesis, such as when research funding is at stake. As for the politicians, they have less incentive to care about supporting cardiac surgery units and far more desire to influence voting patterns among hospital employees.

There is always an abundance of such reasons why professional and political types will ignore each other. No particular skill or energy is needed to list these reasons, although skill and energy are necessary to maintain an awareness of these reasons because they are critical factors in the other's world.

Fear is one of the most important reasons for mutual ignorance. Professionals fear the power of politicians to alter budgets, establish cumbersome regulations, and to abandon health goals and programs that benefit patients. Also, the mysteries of politics can be indecipherable even for the best professional minds. Understanding political forces, if it can be done at all, can prove an interminable dilemma for a scientist who knows the address of each amino acid in a double helix. The temptation is not to enter into the world of politics when it can be ignored without impairing the immediate outcome of research or clinical treatment.

Professionals may observe that politics is immoral, too often requiring the practitioner to abandon principles of good conduct. Still others insist that politics is overwhelmed by opportunism, where expediency washes away scientific reasoning and leaves the public with dubious accomplishments whose glitter, as with fools' gold, quickly fades. For the professional, the politician often embodies themes of tainted power and misapplied mastery. It is power and mastery used for unworthy purposes and for shortsighted goals. Such stereotypes of political and governmental functions often motivate the professional to ignore both.

Similarly stylized depictions can move politicians to ignore health professionals. The latter often appear to inflate the importance of their own work, especially when copies of published reports can be distributed. Politicians have little understanding of the professional literature, except to know that in many cases they have funded research without grasping its full meaning or significance. They also tend to view professionals as greedy, always interested in receiving more dollars for the same work or less. They rarely seem to meet or be impressed by those professionals who work long hours at low pay.

Politicians, like the public they serve, are discouraged when the professional's achievement of individual good seems to be undone by adminstrative or environmental chaos. It is not easy to think of dehumanizing emergency rooms as educational havens for dedicated interns. Neither is there general approval for research such as the 20-year double-blind study of penicillin's effectiveness in treating syphilis, which was reported a few years ago. Another discouragement is the seemingly insatiable desire of professionals for increased autonomy and reduced accountability, both of which are to be granted solely on the grounds of their specialized training and education.

For the political figure, the health professional can be ignored because it appears that the fundamental desire to nurture one's patient is so often corrupted and misdirected by an inability to control those events that affect the patient's life in the long run. The health professional has good intentions but cannot always implement them effectively, as in a hospital. The health professional appears to have only limited mastery and exercises a power dominated by an unrealistic do-gooder philosophy. These factors, for the politician, outweigh whatever positive intentions there may be.

Affiliative Bonding[1] Themes

In contrast to those more frequent times when ignoring each other is the characteristic mode, professionals and politicians do have a need to work together on occasion. When this happens, the themes that served to isolate these groups are often reversed. Politicians who were thought to be too strong, too masterful, and too motivated by objectionable goals are now approached as if they wish to nurture society and accomplish good deeds. All they need, it appears, is a little brain power to point them in the right direction. The health professional will outline a beneficial project that is just as likely to alleviate pain and suffering as it is to win votes. Promises of technological breakthroughs are especially effective in attracting the attention of government officials.

Health initiatives that support the government's desire to provide good quality services also are likely to win approval. When the politician is paying court to the health professional, it is the latter who appears masterful and powerful and the former who wants to play a nurturing role. Ensuing discussions may emphasize the positive aspects of health benefits and diminish any true awareness of their cost and inflationary potential. The professionals are asked for advice on how to make things better. Mastery is attributed to them, perhaps along with a caveat not to overspend. Care must be used, of course, since budget overruns, real or anticipated, provide a way for governmental representatives to discount the power and mastery of the professionals.

A similar set of ambivalent themes is reversed when the professional wishes to attract political attention by raising the classical expectation of a cure in return for some capital or programmatic investment. Again, it would not be unusual for the politician to claim the role of an innocent benefactor who was taken to the cleaners, should the professional fail to produce the promised miracle. A further variation on this theme involves an agreement to put a health professional in charge of some type of clinical or research operation. Even when initial performance expectations are clearly understood, an unexpected outcome may provoke the government's representative. The politician can claim to have erred, been misled, or be subject to new and more urgent priorities, any of which will justify the reassignment or the replacement of the previously annointed professional.

A variety of such scenarios can be imagined, and experienced hands from health and politics are aware of true stories even more vivid than the imagination provides. In each case, it is worth noting that the characteristics attributed to the players when they are ignoring each other are often the reverse of the traits that are attributed when events and needs draw the players to each other. Formerly maligned professionals and politicians are now praised; among those considered weak, strengths are found; where inertia was thought to prevail, action-oriented strategies are discerned.

The assignment of such opposing values—especially along the axes of good-bad, strong-weak, and active-passive—provides the conceptual basis for a stereotyping process which involves health professionals and politicians. Even within the health field, alcohol and drug abuse professionals exchange positive and negative feelings between each other and with their colleagues outside of substance-abuse activities. Politicians from different political parties have the same experience.

The Stereotyping Process

While the public is accustomed to Republicans and Democrats sparring for advantages, they are not as familiar with rivalries within the health professions and only somewhat more aware of conflicts between health professionals and political personnel. Rivalries within the health sector often provoke among laymen a sense of regret about wasted energy and lost opportunities. However, it seems illogical to despair over something so human as rivalry and stereotyping. It has been as much a hallmark of alcohol/drug abuse interactions as it has of professional/political exchanges. Even within the field of alcoholism or drug abuse, rivalries can be strong and persistent.

The work of Charles Pinderhughes,[1] among many, has helped to clarify the underlying forces that account for stereotyping. In suggesting a physiologic basis for stereotyping behavior, Pinderhughes has described a repertoire of processes that support both attachment to as well as disassociation from others. A combination of physiologic, emotional, and intellectual factors powerfully reinforces affiliation or differentiation experiences, thereby contributing mightily to the equilibrium of the psyche. Discriminating and separating behaviors occur more often during periods of stress and psychophysiologic upset, Pinderhughes indicates, since their effect is to calm intrapersonal turmoil and reinforce solidarity in a group of like-minded individuals.

Underlying this emphasis on differentiative versus affiliative experiences is Pinderhughes's concept of "nonpathological paranoia." Such a process helps account for certain facets of stereotyping, such as 1) holding an exaggerated belief when there are obvious chances to look at things realistically, 2) applying the same stereotyped emotional and ideological fervor to patently different situations or persons, and 3) feeling proud of one's rationality at the same time as one adheres unrelentingly to false beliefs. Pinderhughes states, "Data from many sources strongly support the conclusion that human beings are generally paranoid as well as wise. In fact, humans use their intelligence most frequently in support of their various nonpathological paranoias. The paranoias often gain expression in contrived relationships and in contrived social, political, economic, and educational structures.... These paranoias are not accessible to reason, and attempts to alter them lead to confrontations, and often, to violence."[1(36,37)] Ultimately it is worthwhile to have a sense of familiarity with our paranoias, Pinderhughes asserts, because such nonpathological but influential processes are so well integrated into the ways individuals relate to themselves and to the world.

PROFESSIONAL AND POLITICAL SYSTEMS

Three Different Sources of Data Input

As much as Pinderhughes can lead us through the experience of stereotyping and its conflicting themes of differentiative and affiliative bonding, an understanding of the political aspects of alcohol and drug abuse also depends on variables that transcend the level of the individual. While the world of the political person truly can be said to involve all aspects of human life, it is considered in this chapter to include those events and personalities that exist outside of the general health field. Health-related events, regardless of content, are conceptualized here to be in a world unto themselves, with alcohol and drug experience serving as a distinct component within the health world.

Each of these three different strata represents separate sources of data about what goes on in the world. In practice, an individual gathers information daily by collecting news from far more than three sources. However, the purpose here is to examine the manner in which the professional and the political worlds work together, with special reference to that part of the professional world that deals with alcoholism and drug abuse. Each world view—political system, general health system, alcohol and drug abuse treatment system—is considered here to exist in its own right. Also, each world view includes a perspective on the other two components.

Data Sources Acquire Meaning

To the extent that systems, and the events and actors that make them run, are a kind of vocabulary of human accomplishment, the work of Osgood et al[2] provides a method for interpreting significant affective trends as the worlds of politics, health, and substance abuse interact.

Osgood's group has studied extensively the meaning conveyed by human symbols, especially those that comprise our language. As they put it, words are symbols whose meaning occupies a "semantic space" in the minds of their users. Research with semantic differential scales prompted Osgood's team to evaluate the nature of this space. Where several different meanings have been ascribed to a single symbol, they took this as evidence that semantic space has measurable dimensions. They proposed that the three principle dimensions be called evaluation, potency, and activity. These dimensions, respectively, can be defined in casual terms as good versus bad, strong versus weak, and active versus passive. There are other dimensions, of course, and the reader is urged to consult the original texts for further information on this research. However, it is the thrust of Osgood's work to suggest that almost all linguistic symbols convey substantial amounts of affective meaning which can be measured along these three basic axes.

Interaction between the political and the professional spheres can be characterized by a series of symbolic acts and actors capable of stirring the participant and the observer to attribute affective values along at least these three dimensions: active-passive, strong-weak, and good-bad. The impact of these attributed feelings carries through to the alcohol and drug abuse professional and to the politician by arousing either affiliative or differentiative sensations about each other. The attributed effects may also stimulate nonpathological paranoia among both groups, powerfully reinforcing their separatist sentiments. Paradoxically, the identity of both groups is thereby strengthened whenever their intergroup collaborative abilities are undermined. Although cooperative efforts among diversive groups are often sabotaged by such dynamics, it is clearly the opinion of Pinderhughes that such unhappy outcomes are not inevitable. Nonpathological paranoia loses some of its power to disrupt relationships once it is accepted as a familiar occurrence in human interactions.

Pinderhughes and Osgood provide at least two independent, yet relatively congruent, ways of looking at the quality of affective bonding that occurs among politicians and substance-abuse professionals. Their theories allow the participant-observer to speculate more rationally about the motivations for relating that apply to politicians and professionals. These motivations, positive or negative, are as much as collective phenomenon characteristic of either group as they are detectable in the psychodynamics of an individual professional or politician.

Data Input Categorized by Context

Every day, in a variety of settings, events will occur and people will speak up in ways that influence the world of alcohol and drug abuse treatment. Changing circumstances within these settings invariably influence attitudes, especially along the three dimensions cited by Osgood's work: evaluation, potency, and activity. The political significance of such changes can be perceived more clearly by identifying the contextual source of these changes. Recognition of these contexts can either widen or shrink the potential for professional and political systems to work together. Listing them provides a handy reference and an orderly method for observing how political data can influence substance abuse. In systems analysis parlance, the list of context categories shows the kinds of data which have political impact as input to or output from alcohol and drug abuse systems.

Partisan politics Politics based on party affiliation offers a clearcut example. Oddly enough, Republican and Democrat officials may be further alienated by their respective stands on alcoholism and drug abuse. New York State Democrats control the Assembly, and Republicans control the Senate. This has been true for a number of years, a fact that introduces a stability factor which might not be present where the majority shifts more frequently. Furthermore, Republican leadership traditionally has an upstate base, focused on rural and small city areas. Democratic strongholds are primarily the downstate urban areas around New York City plus Buffalo in the west. Such Democratic constituencies have a higher population of drug abusers, both in or out of treatment. This provides the basis for a political schism to function in consonance with an epidemiologic relativism that says that drug abuse is mainly an urban problem, even though the New York City population of alcoholics probably exceeds that of upstate areas.

Measures that promote the treatment of alcoholism have been favored issues for Republicans in New York State, and the so-called alcohol constituency has a tradition of relating well to them. Republicans have become identified as proponents for alcoholism reforms, such as third-party reimbursement for treatment. Drug abuse groups with their urban, Democratic identity ordinarily expect more difficulty in capturing the attention of upstate politicians, regardless of whether they overcome those issues which now keep them from presenting a united front to the public and its elected officials.

Alcohol-drug polarities based on conservative-liberal dichotomies are also noted. This dynamic comes into play because drug abuse is associated in the public mind with criminal activity. Of all the public health and rehabilitative services that government now supports—including mental health; mental retardation; youth services; criminal justice programs, including probation and parole; food stamps; welfare; medicaid and medicare; vocational rehabilitation; temporary job programs; basic health care clinics, including those for venereal disease; and the like—drug abuse treatment can often be portrayed by the ill-informed as the least cost-effective. The public believes that people in treatment continue to rob and steal, break and enter, and assault and rape.

At the federal level, an alcohol-drug dichotomy based on political parties is more difficult to see than has been described for New York

State. This is largely because power in both houses of Congress has until recently rested with a single party for over 20 years. It also is related to the fact that Congress spends small amounts of money on alcohol and drugs, compared to the total federal health-care budget or the total federal budget itself.

Based on personal experience in two other states, South Carolina and Kentucky, the absence of partisan politics as a factor in creating or mollifying any alcohol-drug schism is just as likely to be found at the state level as it is at the federal one. Both South Carolina and Kentucky have largely Democratic legislators and Democratic traditions. Although both states have had Republican governors during the past 20 years, the governors of these two states cannot succeed themselves. This circumstance helps to make them more beholden to the legislative power structure. Furthermore, when a state's governor is from the same political party as the legislative leadership, it is more likely that the two will agree about policy ideas relating to alcohol and drugs. A governor prefers to be in agreement with the legislature in a high number of areas, a factor that enhances the likelihood that all types of major legislation proposed by the executive will be enacted.

On the other hand, when the governor and legislative leaders are from different political parties, they often will take credit at the other's expense. If an alcohol or drug issue can be used to such political advantage, usually it will be. Add to this perspective the probability that debates about alcohol and drug issues, for all the heat they may create among professionals, are usually the political equivalent of a battlefield skirmish rather than a D-Day invasion. Larger issues from the general worlds of politics and health care are where any major political warfare will be waged.

The same is true at the federal level and seems to occur whether or not the President and congressional leadership are from the same party. At one point in his tenure, for example, President Nixon seized the initiative in the drug abuse area by creating a Special Action Office for Drug Abuse Policy (SAODAP). He took the play away from Congress in the public's eye. A few years later, when budget constraints grew, Nixon, Ford, and Carter favored cuts in the fiscal resources of the National Institute on Drug Abuse, deeper cuts than Democratic congressional leaders would allow.

Presidents regularly preempt the Congress by calling for the enactment of unique programs, but this has not been a regular occurrence either in the field of substance abuse or in the related concerns of mental health and mental retardation. It is true that Mrs Rosalyn Carter's special interest in mental health had its parallel with President Kennedy's efforts. Likewise, former HEW Secretary Califano's interest in anti-smoking and anti-alcoholism campaigns are as vivid in the memory of substance-abuse professionals as Nixon's establishment of SAODAP. The point is that executive action in these areas is so infrequent as to be and remain memorable.

In contrast, the major landmarks of the general health field are notable by their diffuse legislative roots. Neither the direct and categorical funding mechanisms of Title XX and Public Law 409 nor the myriad national health insurance programs recently proposed are associated with any particular executive. Nevertheless, in the past few decades every President has conceived of his own version of national health insurance, social security legislation, and federal-state reimbursement plans. Even the establishment of the National Institutes of Health, a premier bureaucratic event, does not carry the imprimatur of a particular President. This reduced visibility for general health goals connotes a certain acceptance by the public and their elected officials regarding the appropriateness of these initiatives, an outcome that many professionals wish held true for substance abuse treatment and research.

Money and economic issues The ability to spend money gives legislators and executive officials their most powerful leverage on alcohol and drug abuse issues. The size of this leverage is usually outside their precise control, however, because economic and monetary variables enlarge or shrink the capacity of money to make a difference in any world.

Spending priorities will vary to the extent that politicians and the public are preoccupied with recession, inflation, energy shortages, unemployment, and international affairs. Alcohol and drug issues almost never have a high priority when these global concerns are activated. From an economic viewpoint, it may make the most sense to campaign for substance-abuse funds when the end of a recession is visible, when unemployment has topped out, and when governmental revenues are about to rise. The likelihood is that substance-abuse expenditures will increase as the economy goes through an improvement cycle. Nevertheless, the need for substance-abuse treatment is continuously present and not subject to the same degree of waxing and waning as is the economy. Many administrators use this fact as a basis for requesting larger budgets for each new fiscal year, although improved productivity of treatment methods sometimes helps to justify reductions in costs.

Economic issues can also be used by politicians to upset commitments that alcohol and drug abuse scientists have already made. Arrangements to evaluate the therapeutic efficacy of long-acting methadone (1-alpha acetylmethadol) were supported by a contractual agreement between a private individual and the National Institute on Drug Abuse when it became apparent in the early 1970s that no pharmaceutical manufacturer wanted to make a financial investment in this project. Upon reviewing the contract in 1979, however, members of Congress and the HEW Secretary became concerned that proper bidding procedures had not been followed to assure fairness of the contract award. This threatened the scientific work with delay and raised the possibility that the results would be viewed with skepticism. It is a clear example of how scientific and political events become related.

Also, in 1979, the *New York Times* reported that a wealthy donor had elected to withhold $5,000,000, or half of his promised gift to a celebrated alcoholism treatment facility, because he did not approve of the program's admission policies. The report suggested that the donor would have preferred for the clinic to treat a greater percentage of employed patients.

These experiences show that money, when given or withdrawn, highlights the role played by alcohol or drug abuse treatment activities in our society. The giving or taking away of money propels these activities into the consciousness of professionals, politicians, and the public alike. Much as advertisements do, these projected images condition the observer's sensitivity and response to alcohol and drug abuse issues. Every new event affords one more opportunity for substance-abuse topics to be evaluated by the observer. The evaluations, of course, may be positive or negative, ambiguous or clearcut. Whatever their valence, the public's opinions become powerful determinants of the context in which alcohol and drug abuse are handled by professionals and politicians.

Personalities The personalities of substance-abuse professionals and politicians are another source of input into the public's consciousness. Modern-day Carrie Nation's, if you will, advocates of substance-abuse causes, share with their temperate predecessor an ability to generate publicity about alcohol and drug abuse by dint of their personal characteristics alone.

Some have come to prominence because of significant achievements in other fields combined with actual personal experience with addiction. This group includes actresses and actors like Mercedes McCambridge and Ben Gazzara, astronaut Edwin Aldrin, sports figures such as Don Newcombe, and Mrs Betty Ford and Billy Carter. Many physicians have achieved acclaim because of their primary interest in substance abuse, such as Morris Chafetz, Robert DuPont, LeClair Bissel, Mitchell Rosenthal, David Smith, and Frank Seixas. A third group includes those who have chosen to use their political or moral authority to reach significant goals on behalf of substance-abuse issues. Examples are Senators Harrison Williams (D-NJ), Donald Riegel (D-Mich), former Senator Richard Hughes (D-Iowa), former HEW Secretary Califano, Monsignor William B. O'Brien, Walt "Clyde" Frazier, Art Linkletter, and Carol Burnett. Senators Williams and Hughes have also acknowledged having had battles with alcoholism. The actions of these and many other prominent persons focus the public's attention on topics of chemical dependence.

The impact of a public personage can be just as diverse as the impact of financial issues or partisan politics. Although many prominent persons advocate substance-abuse control and are viewed with public favor, there are others whose prominence undermines such advocacy. It should not be surprising that alcohol and drug issues frequently become associated with the infamous as well as the famous. As documented in the news media, alleged leaders of organized crime, murderers, rapists, child abusers, muggers, hit-and-run drivers, and criminals of all other types traditionally have been linked, either as abusers or purveyors, to the drug and alcohol fields.

Furthermore, in the entertainment industry, the drug- and alcohol-related exploits of the media stars are often glamorized. Examples of this include cocaine devotees at Studio 54, soap opera characters, in-jokes about marijuana on television talk shows, and "Bonnie and Clyde" movies where alcohol and drugs enhance the dramatic action. These and similar events all play a special complicating role in the public perception of substance-abuse issues. Connecting drugs and alcohol with illegal but socially sanctioned activities seems bound to have political ramifications. The right to use chemicals will be scrutinized closely as society determines whether to interpret conservatively or liberally the individual freedoms that are guaranteed by the Constitution.

Legislative and regulatory decisions While the role of lawmakers is more extensively documented in another chapter, it is important to include here a comment about the input of legislative and regulatory decisions on clinical practice. Typically, laws are developed by elected officials, whereas regulations are more likely to be the work of civil servants and appointed leaders. Legislation characteristically involves global concepts, such as prohibitions against the use of substances, licensing of those permitted to sell or dispense substances, and penalties for violations of those statutes. Regulations provide a detailed, structured set of guidelines for practitioners of the therapeutic arts, although rarely are they so contemporary that they can function as standards of care to which everyone is expected to adhere.

Restated from a different perspective, standards of care are usually established by those institutions that deliver treatment services, while regulations provide guidelines within which the standards may be set. Legislation supplies the legal basis for the government's right to intrude into the health-care delivery process by setting regulatory guidelines. The Joint Commission on the Accreditation of Hospitals, for example, sets standards, while a State Department of Health mandates regulatory compliance of licensed hospitals. Standards may require a richer staff/patient ratio, for example, compared to the minimum ratio stipulated in a regulation. As a further distinction, a law often provides that the authority to set a minimum staff-to-patient ratio will be vested in a particular state (or federal) agency.

The tripartite functions of legislation, regulation, and standard setting appear to leave little room for clinical initiative in the alcohol and drug areas, or in any other medical area, for that matter. However, the reality is that a wide variety of clinical practices will flourish despite these apparent restrictions. Methadone treatment regulations, perhaps the most stringent of all governmental intrusions into clinical practice, still permit different approaches to be taken regarding such crucial decisions as the tempo by which a patient approaches abstinence, the awarding of take-home privileges, and even the initial admission into treatment. Likewise, alcoholism treatment varies widely depending on who is in charge. For example, some experts favor regular use of adjunctive medications during alcohol withdrawal, and others work to avoid them at all costs.

A number of clinicians tend to view legislative and regulatory pressures as violating a professional right to autonomy. The public and their elected representatives traditionally see regulations and laws as protections against unscrupulous and incompetent practitioners.

Special interest groups, organizations, and agencies Lobbyists will want to influence legislative and regulatory actions in all but the most mundane or routine situations. In health affairs, official groups such as the American Medical Association or the American Hospital Association are among the better recognized lobbying organizations. In a land where free speech reigns, anyone can assume the trappings of a lobbyist, and a number of individuals and groups have assumed this role in more or less visible ways. Groups such as the National Council on Alcoholism (NCA), the National Drug Congress (NDC), and the Association for Medical Education and Research in Substance Abuse (AMERSA) are among those that act as spokesmen for substance-abuse issues. Other groups, such as Alcoholics Anonymous, specifically avoid any link with government or outside funding sources so as to preserve their clinical autonomy.

Private, not-for-profit groups are supported mainly by volunteer contributions of time and money, an organizational trait that permits them to say they have no vested interest in governmental decisions or scientific research proposals. However, individual members certainly are not prohibited from membership if they have a vested interest in political actions. Private groups within the substance abuse field include NADAP, the National Association on Drug Abuse Problems, NCA, and AMERSA.

Many advocacy groups represent clients, health-care provider agencies and their employees, labor unions, and professionals who benefit or suffer directly when legislation or regulations are enacted or when government funds are spent. Some of these are not-for-profit voluntary organizations, and some are not. In the field of substance abuse specifically, the groups include the Alcohol and Drug Problems Association (ADPA), the National Association of State Alcohol and Drug Abuse Directors (NASADAD), Therapeutic Communities of America (TCA), the National Drug Abuse Conference (NDAC), and the National Association of Puerto Rican Drug Abuse Programs, among others. Alcohol-related groups include the American Medical Society for Alcoholism (AMSA) and the American Labor-Management Association to Control Alcoholism (ALMACA).

Even the official federal and state agencies designated by political forces to act on behalf of substance-abuse issues will exert pressure on the legislative process. Unlike private advocacy groups, these agencies are under the direct control of political representatives. Agency policies and budgets can be influenced by these representatives whenever substance-abuse problems receive public notice. While this may seem to professional eyes to be an all too expedient mechanism for resolving tough clinical dilemmas, it does provide an indirect but influential means for the public to have some say about how and what kinds of services will be provided. At the federal level, these agencies include FDA, NIAAA, NIDA, DEA, and to a more limited extent the Treasury Department, the Veterans Administration, other sections of the Department of Health and Human Services, and portions of the Departments of Defense and Education.

The need and desire for public participation in health care decisions also is reflected in the increasingly complex review mechanisms provided by federal legislation in the past decade. Regionalized health systems agencies now assure that federal dollars are not spent on health care delivery without local citizen review. Health systems agencies must also certify that a need for new services or for changes in old services exists before federal funds can be spent to respond to those needs. These review mechanisms supplement the more direct and more immediate influence that legislators and governors can exercise when they modify the budgets of substance-abuse agencies.

Yet another way in which legislators, governors, and other officials may influence the delivery of health care is by auditing the financial and programmatic actions of an agency that either administers or receives public funds. The potential embarrassment to professionals and politicians once the findings of an audit are released is well recognized as a motivating force whenever the expenditure of government funds is involved. The same can be said for program reviews conducted by legislative fact-finding committees. In recent years, the United States House of Representatives Select Committee on

Narcotic Abuse and Control, headed by Lester L. Wolff (D-NY), has been an active proponent of various drug abuse reforms.

Social, cultural, and demographic factors Political and professional people must also be alert to the variety of social, cultural, and geographic factors that intertwine with the field of substance abuse. While these factors are so complex and varied that they are catalogued more extensively in other chapters of this book, it is important to keep in mind the manner in which substance-abuse issues can be influenced by such politically important items as ethnicity, religion, economic class, education, and similar variables. Furthermore, what works in New York may not apply in South Carolina or Kentucky, and vice versa. Also, the traditions of the United States should not be routinely applied in other countries. While it is obvious to make these points, making these points does not obviate the professional and the politician from informing each other about one's own world and the variables that make a difference in those worlds.

Technology Another variable that transcends the political-professional dichotomy is the influence of scientific technology. There is great power, in the attention getting sense of that word, attributed to scientific breakthroughs which provide either the appearance or the fact of revolutionizing medical care. The discovery of penicillin and the development of computerized axial tomography are two successful examples.

The field of substance abuse could very well be similarly influenced by some development that markedly simplifies the public's understanding of addiction or the patient's willingness to be treated. At one time it was hoped that methadone would prove to be a panacea for those addicted to heroin, much as heroin was at one time viewed as the answer to morphine addiction. A recent development, this one at a basic science level, is the discovery of the endorphins. Knowledge about these neurologically active polypeptides may revolutionize our comprehension of how the body perceives pain and develops tolerance to analgesics. Advances in either of these areas would bring swift changes to the relationships of politicians and substance-abuse professionals.

The point could be made that whatever these changes would be, professional-political rapport would change in directions that already have been established. The same stereotypes would apply and the same dichotomies would exist, but the new technology would supply an opportunity to approach these differences from a new perspective. Plus ça change, plus c'est la même chose.[3]

THE TASKS OF PROFESSIONAL AND POLITICAL LEADERSHIP

To the extent that events in the world of alcoholism and drug abuse can be influenced by politics, it seems worthwhile to focus somewhat more extensively on leadership activity. Actually, the leader's dilemmas and decision points confront all of those who work in the substance-abuse disciplines. A perspective on leadership in this field need not be considered narrowly important.

Leaders in politics and the professions, much like leaders in most fields, have the capacity to lay aside their differences and skepticism as they search for a common ground. I am reminded of a time when, as a deputy commissioner, I had invited to my office the head of a large treatment program, wanting to take him to task for operating one of his clinics in a way that was counter to the goals of my agency. It was in both of our best interests to work together cooperatively, yet there were no legal requirements that he base his decision on state agency policies any more than the state needed to be shy about listing its objectives and asserting its willingness to use fiscal or administrative judgments to influence the program leader's behavior.

As our conversation proceeded, it seemed more appropriate to quell the desire to argue and instead to search cooperatively for a solution to a problem that neither of us alone could ease. It is worth remarking, too, that both of us were experienced enough to take the facts of our problem situation and arrange them in such a way as to make either side look like the good guy and the other, the bad.

Leadership involves the ability to see both sides and to be an advocate for one side or another. It also involves the capacity to know when advocacy can be overdone. In the alcohol and drug abuse field, the role of advocate is only one of many roles available to the leader, just as choosing a role is only one of at least three kinds of choices a leader has to make.

Select a Level of Achievement

The scope of a project, its bigness or smallness, the number of people who will be affected by a decision, and the amount of money to be spent are all variables that describe in a tangential way what a leader must evaluate when choosing a "level of achievement."

Also included in this concept are the required craft, skill, strategy, and subtlety. Notable achievements need not be thought of as large only, such as the acquisition of a sizable federal grant; nor elegant only, such as a double-blind, cross-over study of an agent to ease narcotic withdrawal syndromes; nor theoretically consistent only, such as a drug-free clinic that decides not to use methadone to detoxify new clients versus one that does use methadone. Level of achievement refers to the complexity and simplicity of the leader's work. It embodies the notion that complex and simple can coexist.

In systems analysis terminology, level of achievement can refer to the leader's decision about which organizational components are to be used to identify or accomplish goals. An alcohol program executive may decide to focus on making changes in the lives of patients, staff, the board of directors, or the community, the media, state government, or the entire nation. This executive will select which of these organizational levels to recruit into an action plan, and whether this recruitment will be temporary or permanent.

Another example is the drug treatment executive who is forced to choose between vocational and housing needs and the basic group of medical and counseling services that the clientele require. Leaders, at certain times, will find it useful to expand the universe of their concern, while at other times they may prefer to encapsulate a single issue, making it their own and seeming to ignore the manner in which a solitary issue relates to the others.

Level of achievement choices also include the selection of contextual categories within which the leader's agency can generate data output to highlight and to facilitate objectives. One measure of accomplishment will be the leader's ability to operate within such varied contexts as partisan politics, fiscal and economic trends, the personalities of notable figures, legislative and regulatory environments, and the diverse forces of special interest groups.

Finally, and it needs to be said again, levels of achievement imply no connotation of an inherent preference for higher versus lower levels, bigger versus smaller levels, or good versus bad ones. Achievement level is, in some ways, like beauty, being what the eye of the beholder wishes to see. One leader's high level may be seen as low by another, and so on. Level of achievement is a dimension that is independent of valuation. Recently, an automobile bumper sticker has appeared which declares with some professional hubris, "If you can read this, thank a teacher." Being able to read is a level of achievement, here separated by a comma from the valuation that the driver assigns to it.

Leaders generally see this kind of separation when they sit down with one another. They can put aside cynicism and competitiveness while they search for ways in which to cooperate. Cooperation in this setting refers to how two or more leaders will value one another's work, possibly providing a basis for joint actions. A leader's day or week should be structured in such a way that such experiences will occur with some regularity. Still, it would be foolhardy to believe that a leader's public behavior totally would eschew competitiveness and stereotyping. Even Ghandi had to have the British to fight.

Select Roles that Permit Achievement

Choosing among many roles to play, in public and in private, is the second major selection task that a leader encounters. Ghandi, of course, is a fine example of a person who became stereotyped by the roles he played. For the alcohol or drug abuse professional, he is a model leader from the world of politics who, as a minority representative, projected himself into the consciousness of the majority.

Although stereotyped, a leader is not necessarily locked into one single role to play. Advocacy for a cause has as its opposite the role of accommodation. The innovator shares the limelight with the champions of the status quo. Those leaders whose role is playing to a well-established constituency will stand in contrast to others who are working to create new constituents. Sometimes the same leader will try to do both. Leaders who project themselves as doers and actors will be seen differently from the leaders who function as teachers and advisers. Some leaders prefer to do the work themselves; in other styles of leadership, gratification comes vicariously because the work of others has been facilitated by the leader. Each of these roles is almost continuously available to every single leader, consistent with the leader's own personality and the work environment.

Circumstances sometimes force a change in roles. Upon taking a new job, a leader who espouses the use of minor tranquilizers in the treatment of alcohol withdrawal syndrome may come to represent a larger, broader constituency where the drug-free treatment of this syndrome must be allowed to exist and be evaluated objectively. Whether in a new city or within the walls of a long-familiar institution, job changes of any kind are a common way for persons interested in leadership to develop the necessary skills for making choices about their levels of achievement and the roles they will play. Change forces political or professional people to encounter and to engender new constituencies for themselves as well as for those aspects of alcoholism and drug abuse that have their attention.

Know the Value of Leaders

Career development is one of the more obvious means for confronting the leader with a third kind of choice, the choice of values. Leaders have value (speaking collectively about them as a group), and leaders also have values (speaking of them as individuals).

The value of leaders has been asserted by the constructs of many traditional and familiar analytic methods. Leadership functions in the systems analysis view are described by the term "decider." In Broskowski's words, "the decider subsystem is used for system control and integration. The decider receives information from all other subsystems, including internal and environmental sources.... [It] transmits information codes to other subsystems, including those that process matter and raw energy. The decider thereby controls the entire system.... [It] is easily recognized as the executive or boss."[4]

Freudian views of leaders are derived from the conflict resolution activities of individuals, all of whom must contend with instinctive forces within the psyche versus the external requirements of culture and society. Leadership behaviors are seen as one way an individual chooses to resolve anxieties that stem from these conflicts. For Jung, the collective aspects of the unconscious are determining factors. Archetypal forces shape personal behavior and lend a cultural and symbolic value to individual acts. Helpful as these views are for an in-depth appraisal of leadership, they suggest only indirectly the actual values that a leader may be called upon to espouse. The field of alcohol and drug abuse is replete with difficult value choices, none of them entirely limited to either the alcoholism or the drug abuse field.

Know a Leader's Values

Perhaps the most long-standing ethical dilemma in the substance-abuse area is whether alcoholics and drug abusers will have access to health care. Substance abusers as patients are typically viewed with disdain by the health professions, with a concomitant denial of services that is either purposeful or inadvertent. In contrast, those impaired patients who enjoy sufficient wealth, education, or privilege are far more likely to get treatment for their substance-abuse syndromes, although even their chances of receiving care from an uninformed professional remain high. Leaders in this field today are constantly faced with decisions that relate to the issue of equal access to health care services for the drug abuser and the alcoholic.

Another basic value concerns a person's freedom to choose a way of living and a way of dying, in contrast to the more conservative position that life must be preserved at all costs. Choosing among these values is an obvious task when a patient who is about to be discharged threatens to drink once again or to take more drugs right after having been saved from a life-threatening episode. A leader can easily demur to the clinician in such circumstances.

However, a leader's choices concerning this value are decidedly more complex when dealing with the topic of substance-abuse prevention. The majority of persons in America prefer to take risks drinking, smoking, and using drugs. Leadership behavior may call for a reasoned espousal of the right to take risks as well as recognition of the popular truism that the greatest risk may be to take no risks. One practical way in which this impacts leaders and all their colleagues is deciding how to induce the public to accept without prejudice the fact that relapses occur in most substance-abuse syndromes, and that quick and permanent cures are the rare exception instead of the rule. The public will want to know whether a leader believes patients are to be afforded the right to take risks with their lives again, especially if these risks undermine treatment that was paid for with public funds.

A third value which the substance-abuse leader encounters is the public scrutiny of personal behavior versus the right to privacy. The leader confronts this value conflict in many ways. There is the question of the leader's own use of alcohol, drugs, and other substances. The leader is often called on to decide whether to ask others to do as I do, or do as I say. Another type of situation involves the confidentiality of a patient's identity versus the desire to provide the best service to the patient. Disclosure of a patient's circumstances to another professional or to another leader may often seem the most expedient way to arrange help for a client. This process is governed by legal statute, but its implementation can get complicated when details of the patient's situation are used to impress others with the leader's own accomplishments, or when requested details are withheld to inflate the leader's sense of importance. A third variation of this ethical difficulty involves the transfer of funds, with the possibility that leaders can be thought of as using their position to benefit themselves or their institution at the undue expense of a patient or the public.

Finally, the leader can choose to contend with or to ignore questions about drug and alcohol abuse and the quality of life. Related to this are decisions about who determines how to measure this quality. The First Amendment guarantees the right of free speech in the United States, yet newspaper and television stations cannot report all the details and subtleties that surround substance-abuse issues. A leader in the field may decide that his legitimate role includes the right to comment on how well the media handles their responsibility towards publicizing substance-abuse issues. Julio Martinez, for example, shortly after he was appointed Director of the New York State Division of Substance Abuse Services, called a press conference to criticize the media for glamorizing substance abuse and making media stars of those who dabbled in drug abuse. Clearly, leaders in the drug and alcohol field have this option.

In summary, this chapter has reviewed the political dimensions of alcoholism and drug abuse by discussing the stereotyping process that exists between professionals and politicians and by listing and describing those contextual variables that typically influence the dialogue between politicians and professionals. Special attention was given to the variable of leadership and to the observation that leaders ordinarily are involved in selecting levels of achievement, roles, and values.

REFERENCES

1. Pinderhughes CA: Differential bonding: Toward a psychophysiological theory of stereotyping. *Am J Psychiatry* 136:33–37, 1967.

2. Osgood CE, Suci JG, Tannenbaum PH: *The Measurement of Meaning.* Chicago, University of Illinois Press, 1957.

3. Karr A: *Les Guepes.* 1849.

4. Broskowski A: Application of general systems theory in the assessment of community needs, in Bell RA, Sundel M, Aponte JF, et al (eds): *Need Assessment in Health and Human Services.* Louisville, University of Louisville School of Medicine, Department of Psychiatry, 1976, p 64.

ADDITIONAL READINGS

Allport GW: *The Nature of Prejudice.* Cambridge, Mass, Addison-Wesley, 1979.

Brown ER: *Rockefeller Medicine Men: Medicine and Capitalism in America.* Berkeley and Los Angeles: University of California Press, 1979.

Carlson RJ (ed): *Future Directions in Health Care: A New Public Policy.* New York, Ballinger, 1978.

Deadalus: "Doing Better and Feeling Worse: Health in the United States," Volume 106, Winter, 1977. The entire issue has relevance to the political dimensions of the health field. Special attention is merited by the articles of John H. Knowles, Renee C. Fox, and Aaron Wildavsky.

Freud S: *Character and Culture.* New York, Collier, 1963.

Freud S: *Civilization and Its Discontents.* London, Hogarth, 1955.

Illich I: *Medical Nemesis: The Expropriation of Health.* London, Pantheon, 1975.

Jung CG: *The Archetypes and the Collective Unconscious,* ed 2. Princeton, NJ, Princeton University Press, 1969.

Kitrie NN: *The Right to be Different. Deviance and Enforced Therapy.* Baltimore, Johns Hopkins University Press, 1971.

Miller JG: *Living Systems.* New York, McGraw-Hill, 1977.

INDEX